SMART GRID SYSTEMS

Modeling and Control

SMART GRID SYSTEMS

Modeling and Control

Edited by

N. Ramesh Babu, PhD

Apple Academic Press Inc.
3333 Mistwell Crescent
Oakville, ON L6L 0A2
Canada

Apple Academic Press Inc.
9 Spinnaker Way
Waretown, NJ 08758
USA

© 2019 by Apple Academic Press, Inc.

First issued in paperback 2021

Exclusive worldwide distribution by CRC Press, a member of Taylor & Francis Group
No claim to original U.S. Government works

ISBN 13: 978-1-77-463065-5 (pbk)
ISBN 13: 978-1-77-188625-3 (hbk)

Library and Archives Canada Cataloguing in Publication

Smart grid systems : modeling and control / edited by N. Ramesh Babu, PhD.

Includes bibliographical references and index.
Issued in print and electronic formats.
ISBN 978-1-77188-625-3 (hardcover).--ISBN 978-1-315-11112-4 (PDF)

1. Smart power grids. I. Ramesh Babu, N., editor

| TK3105.S63 2018 | 621.31 | C2018-901675-2 | C2018-901676-0 |

CIP data on file with US Library of Congress

ABOUT THE EDITOR

N. Ramesh Babu, PhD, is currently working with M. Kumarasamy College of Engineering, Karur, India. He has over 15 years of teaching experience with 8 years of research experience. He has chaired or organized many conferences, workshops, and faculty development programs and has been the keynote speaker, session chair, and technical program member. He is an active reviewer for many international journals, including those published by Elsevier, IEEE, and IET. He is serving as Associate Editor for the *IEEE Access Journal* and is an editorial board member for several other journals. He has published several technical papers in national and international conferences and international journals with high impact factors.

Dr. Babu has guided many undergraduate and postgraduate student projects and is presently guiding many such projects. Under his mentorship many students won competitions for their projects, designs, and papers. He is a member of several professional organizations such as IEEE, IAENG, and ACDOS. His current research includes wind speed forecasting, optimal control of wind energy conversion systems, solar energy, and soft computing techniques applied to electrical engineering.

CONTENTS

LIST OF CONTRIBUTORS

P. Arulmozhivarman
School of Electrical Engineering, VIT University, Vellore, India

N. Ramesh Babu
M. Kumarasamy College of Engineering, Karur, India

Amjed Hina Fathima
School of Electrical Engineering, VIT University, Vellore, India

A. Rini Ann Jerin
School of Electrical Engineering, VIT University, Vellore, India

K. Kumar
School of Electrical Engineering, VIT University, Vellore, India

K. Palanisamy
Department of Energy and Power Electronics, School of Electrical Engineering, VIT University, Vellore, India

N. Prabaharan
School of Electrical Engineering, VIT University, Vellore, India

J. Prasanth Ram
School of Electrical Engineering, VIT University, Vellore, India

N. Rajasekar
School of Electrical Engineering, VIT University, Vellore, India

V. Ramesh
School of Electrical Engineering, VIT University, Vellore, India

Sarat Kumar Sahoo
Department of Electrical Engineering, School of Electrical Engineering, VIT University, Vellore, India

G. Saminathan
School of Electrical Engineering, VIT University, Vellore, India

S. Saravanan
School of Electrical Engineering, VIT University, Vellore, India

V. Sridhar
School of Electrical Engineering, VIT University, Vellore, India

Ramji Tiwari
School of Electrical Engineering, VIT University, Vellore, India

S. Umashankar
School of Electrical Engineering, VIT University, Vellore, India

K. Vikram
School of Electrical Engineering, VIT University, Vellore, India

LIST OF ABBREVIATIONS

AC	air conditioning
ACF	autocorrelation function
AI	artificial intelligence
AIC	Akaike's information criteria
AMI	advance metering infrastructure
AMR	automatic meter reading
ANFIS	adaptive neuro fuzzy inference system
ANN	artificial neural network
AR	auto-regressive
ARIMA	auto-regressive integrated moving average
ARMA	auto-regressive moving average
ARMAX	auto-regressive moving average model
BEMS	building energy management systems
BESCOM	Bangalore Electricity Supply Company
BESS	battery energy storage system
BIC	Bayesian information criteria
BPL	broadband power line
BPN	backpropagation networks
CAES	compressed air energy storage
CCT	clean coal technology
CHB	cascaded H-bridge
CHP	combustion heat to power
CPP	critical peak pricing
DA	distribution automation
DCU	data concentrator units
DER	distributed energy resource
DFIG	doubly fed induction generator
DMS	distribution management system
DPC	direct power controller
DR	demand response
DRM	demand and response management
DSM	demand side management
DTC	direct torque control

DVR	dynamic voltage restorer
ECS	energy conservation system
EMS	energy management system
EPRI	Electric Power Research Institute
ESS	energy storage system
ESSS	exponential smoothing state space
ETP	European Technology Platform
EV	electric vehicle
FACTS	flexible AC transmission system
FAN	field area networks
FC	fuel cell
FCL	fault current limiters
FERC	Federal Energy Regulatory Commission
FESS	flow battery energy storage systems
FLC	fuzzy logic control
FOC	field oriented control
FRT	fault ride through
FWD	free-wheeling diodes
GA	genetic algorithm
GBPS	giga bits per seconds
GEA	Geothermal Energy Association
GeSC	generator side converter
GHG	greenhouse gas
GHI	global horizontal irradiance
GIS	Geographic Information System
GPM	gallons per minute
GPRS	General Packet Radio Service
GPS	Global Positioning System
GSC	grid side converter
GSI	grid side inverter
HAN	home area network
HCS	hill-climb search
HEMS	home energy management systems
HESS	hybrid energy storage systems
HEV	hybrid electric vehicles
HIL	hardware in-loop
HRE	hybrid renewable energy
HRES	hybrid renewable energy systems

IAP	interoperability architectural perspective
ICT	information and communication technology
IEC	International Electrotechnical Commission
IED	Intelligent Electronic Device
IEEE	Institute of Electrical and Electronics Engineers
IEGC	Indian Electricity Grid Code
IGBT	insulated-gate bipolar transistor
IHD	in-home devices
INC	incremental conductance
IoT	Internet of things
IP/MPLS	Internet protocol/multi-protocol label switching
IWGC	Indian Wind Grid Code
KBPS	kilobits per second
LIDAR	light detecting and ranging
LM	Levenberg–Marquardt
LQG	linear quadratic Gaussian
LTE	long-term evaluation
LVRT	low voltage ride through
MA	moving average
MAE	mean average error
MAPE	mean average percentage error
Mbps	megabits per second
MDMS	Meter Data Management System
MERS	Magnetic Energy Storage Device
MPP	maximum power point
MPPT	maximum power point tracking
MSE	mean square error
NAN	neighbor area network
NARNN	nonlinear auto-regressive neural network
NFC	near-field communication
NIST	National Institute for Standards and Technologies
NN	neural network
NPC	neutral point clamped
NWP	numerical weather prediction
OLS	orthogonal least square
OMS	outage management system
ORM	outage and restoration management
OTC	optimal torque control

P&O	Perturb & Observe
PACF	partial auto-correlation function
PAP	Priority Action Plan
PCC	Point of Common Coupling
PE	power electronic
PEC	power electronics converter
PHEV	plug-in hybrid electric vehicles
PHS	pumped hydro storage
PI	proportional integral
PID	proportional-integral- derivative
PLC	power line communication
PLL	phase locked loop
PMSG	permanent magnet synchronous generator
PMU	phasor measurement unit
PONs	passive optical networks
PR	proportional-resonant
PSB	polysulphide–bromide flow battery
PSC	partial shaded conditions
PSF	power signal feedback
PV	photo-voltaic
PWM	pulse width modulation
RBFN	radial basis function networks
RES	renewable energy source
RFID	radio frequency identification
RMS	root mean square
RMSE	root mean square error
RSSI	received signal strength indicator
RTP	real-time pricing
SARIMA	seasonal auto-regressive integrated moving average
SC	super capacitor
SCADA	Supervisory Control and Data Acquisition
SCES	super capacitors or electric double layer capacitors
SCIG	squirrel cage induction generator
SDR	series dynamic breaker
SFCL	superconducting fault current limiter
SG	smart grid
SGIP	Smart Grid Interoperability Panel
SGSC	series grid side converter

SMC	sliding mode controller
SMES	superconducting magnet energy storage
SOC	state of charging
SOM	self-organizing maps
SSE	sum of squared error
SSFCL	solid-state fault current limiter
SSSC	static synchronous series compensator
STATCOM	static compensator
STC	standard test conditions
STFCL	switch type fault current limiter
SUN	wireless smart metering utility network
TDH	total dynamic head
THD	total harmonic distortion
TOU	time-of-use
TSI	total sky imager
TSR	tip speed ratio
UHV	ultra high voltage
UVRT	under voltage ride through
V2G	vehicle to grid
VOC	voltage oriented controller
VPN	virtual private network
VPP	virtual power plant
VRB	vanadium redox-flow battery
WAMPAC	wide area monitoring, protection and control
WAMS	wide area monitoring system
WECS	Wind Energy Conversion System
WiFi	wireless fidelity
WiMAX	Worldwide Interoperability for Microwave Access
ZBB	zinc–bromine flow battery
ZCS	zero current switching
ZVS	zero voltage switching
ZVT	zero voltage transition

PREFACE

Electric power systems are being transformed from older grids to smart grids across the globe. The main aim of this transition is to meet the current needs such as reducing carbon footprints, finding alternate decaying fossil fuel, eradicate the losses in the available system, and introduce the latest technologies of information and communication technologies (ICT). Smart grid development advances drastically along with the continuing growth of renewable energy technologies, especially wind and solar power, the growth of electric vehicle, and the huge demand for electricity.

The aim of this book is to introduce the smart grid and provide a basic understanding. The volume also focuses on recent technological advancements in smart grids. The book is organized as below:

Chapter 1 provides an overview of the smart grid, along with its needs, benefits, challenges, existing structure, and possible future technologies.

Chapter 2 discusses solar photovoltaic (PV) system modeling and control along with battery storage, which is a part of smart grids.

Chapter 3 covers the issues and challenges of wind turbines connected to the grid and also discusses the various advances in the fault ride through capability with solutions.

Chapter 4 discusses the control strategies for renewable energy systems. In this chapter, control strategies for solar PV, wind, and the hybrid systems are described in detail.

Chapter 5 focuses on power electronic converters, which are used to improve the characteristic of renewable power generation to match with the grid capability. This chapter details various topologies that have been developed recently for the grid applications.

Chapter 6 describes the inverter topologies adopted in integrating the renewable power generated to the grid. The chapter also compares various inverter configurations suitable for microgrids.

Chapter 7 deals with the basics of an energy storage system and its need for microgrids. A case study with a hybrid storage system is discussed in detail.

Chapter 8 describes various forecast techniques for renewable energy systems such as solar and wind. This chapter provides an in-depth understanding of various forecast algorithms with a case study and comparison of different forecasting techniques.

Chapter 9 covers the basics and structure of the energy management system in a smart grid. This chapter includes advanced metering, various communication protocols, and the cyber security challenges in the smart grid.

Chapter 10 focuses on the electric vehicle technology and its interaction with a smart grid. The impact of integrating electric vehicles with smart grids has been detailed along with the potential impact and challenges.

Finally, Chapter 11 discusses various challenges and research perspectives in smart grids. An overview of the current status of smart grids in different countries are discussed along with the challenges and benefits.

The editor and the contributing authors hope readers will benefit and gain a basic understanding of the smart grid and its control. This book will definitely enhance the knowledge of readers and make an impact in realizing smart grids worldwide in the near future.

CHAPTER 1

SMART GRID OVERVIEW

SAMINATHAN GANESAN, V. RAMESH, and S. UMASHANKAR[*]

School of Electrical Engineering, VIT University, Vellore, Tamil Nadu, India

[*]*Corresponding author. E-mail:umashankar.s@vit.ac.in*

CONTENTS

ABSTRACT

This chapter provides an overview of the smart grid concept, and the factors driving its need. The limitations of present-day power system infrastructure are presented and also, it explains briefly how the technological advancement help in realizing smart grid in the real world along with the associated challenges and benefits.

1.1 INTRODUCTION

Recent technology advancement and day by day growing energy demand necessitates significant role for smart grid. The concept of smart grid exists for quite some time, but there are only countable smart grid installations mainly in the academic institutions and far remote locations. Until the arrival of Internet of Things (IoT), smart grid's terminology was niche technology. The present communication technology advancement has revolutionized the electrical energy generation and distribution. Smart grids with more renewable source penetration are poised to become intrinsic aspect of modern power system. The new technology advancement (Kouro et al., 2010) will help consumer and service provider to take absolute control over cost; reliability and energy sustainability also enables active participation of all stakeholders.

The smart grid is a combination of hardware, management, and reporting software, built atop an intelligent communications infrastructure. In the world of the smart grid, consumers and utility companies alike have tools to manage, monitor, and respond to energy issues. The flow of electricity from utility to consumer becomes a two-way conversation, saving consumers money, energy, delivering more transparency in terms of end user use, and reducing carbon emissions.

The total concept of smart grid can be devised into four main sections: (1) distributed energy production (Chowdhury et al., 2009) by means of renewable energy sources locally, (2) efficient and cost-effective energy management (Gaurav et al., 2015) system, (3) managing intermittency nature of renewable sources by using storage systems, and (4) managing intelligent control and communication system for decision-making and execution. These four elements are capable of bringing paradigm shift to the present day energy system. Figure 1.1 shows the progression of smart grid in current scenario.

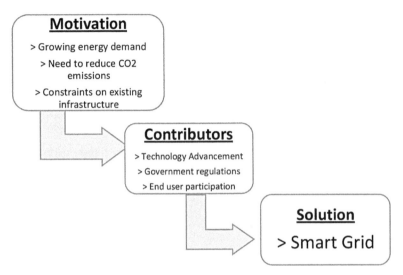

FIGURE 1.1   The progression of a smart grid solution.

1.2   NEED FOR SMART GRIDS

In today's world scenario, stringent carbon emissions are mandated post COP21 agreement deployment across the globe. For achieving super economic growth under these circumstances, the evolution of energy industry is important. A smart grid system with a technology intensive and superior communication network will enable locally controlled and highly reliable power. Depleting fossil fuel reserves will make smart grids with higher renewable energy sources penetration increasingly cost-effective. With the combination of renewable (Teodorescu et al., 2011) and storage systems (Trowler et al., 2012; Suct et al., 2009), the peak hour demand can be well managed, hence reducing cost of energy during peak hours of consumption. The smart grid will manage demand and supply to meet creatively at all points of time, by using storage and high-cost instantaneous power sources at the local level. At neighborhood, district, state, and national levels, this will reduce the capital cost on installed capacity.

The recent study by *The Wall Street Journal* revealed that any assault on just nine key substations among the total of 55,000 could paralyze the entire United States power system for many weeks to months. So in order to prevent such eventuality and to improve the reliability of power system, localized generation, distribution, and consumption would be the right pick. In nutshell, the smart grid solution will fulfill environmental, reliability, sustainable energy, and economic growth requirements. Over the past few years, the smart grid has been developing quickly, but unevenly. The recent events such as Fukushima nuclear reactor blast in Japan have made the energy equation more complex everywhere, further accelerating the need for smart grid with increased renewable penetration (Omran et al., 2009).

Figure 1.2 shows the growth of renewable energy penetration into total energy generation over next three decades across the world. The Europe will take the maximum share of 44% of renewable generation in their total energy production.

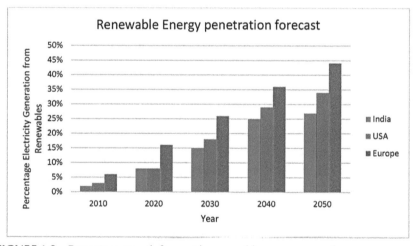

FIGURE 1.2   Percentage growth forecast in renewable energy generation.

1.3   SOCIETAL BENEFITS

- Delivers high quality power locally
- Easily controllable to meet local dynamics
- Secured and reliable operation
- Optimize consumption pattern based on cost of energy
- Energy buying and selling with utility with real-time data analysis
- Helps to improve environment through green energy
- Reduced line losses, thereby reduction in cost

- Customized operation to the distinct needs (e.g., military and space applications)

1.4 CHALLENGES

- The prevailing energy policies of many countries are not flexible or favorable for localized generation and distribution.
- Lack of regulation for real-time pricing and buying and selling by consumers.
- Successful business formula/model yet to be established.
- Complexity in interconnection with utility grid.
- Interoperability of various sources and source management controllers.

1.5 OUTLOOK ON EXISTING POWER SYSTEM INFRASTRUCTURE

With the existing generation, transmission, and distribution infrastructure, to supply 1 MW power at the consumer end, the actual generation to be 1.2 MW to meet the line losses, transformation losses and congestion losses. This requires the installation capacity of 2.2 MW considering 45% of average fleet capacity factor. On the contrary, in case of distributed generation of 1 MW, the install capacity requirement is only 1.4 MW.

Table 1.1 lists the average power outage index; as per this data, 106 min of average black out in a year would cost $80B–150B revenue (Data source: Lawrence Berkeley National Lab, IEEE 1366, EIA data EPRI).

TABLE 1.1   Electric Power System Reliability Data of United States.

Sr. no.	Year	Average power system interruption in minutes	Remarks
1	2000	100	–
2	2005	106	–
3	2010	120	–
4	2015	135	–
5	2020	145	Extrapolated
6	2025	158	Extrapolated

Also, it is predicted that the cost of electricity from fossil fuel is expected to increase 7% by year 2020 and 15% by year 2030 from now.

The natural gas cost is expected to increase by 56% by 2020 and 89% by 2030. The existing electrical infrastructure in most of the developed nations is century old; this aging grid is inflexible for the expansion and mandates a massive transition in energy ecosystem. The ideal solution would be smart grid. Figure 1.3 shows the priorities of energy infrastructure and consumer requirements across different geography and its solution through smart grid.

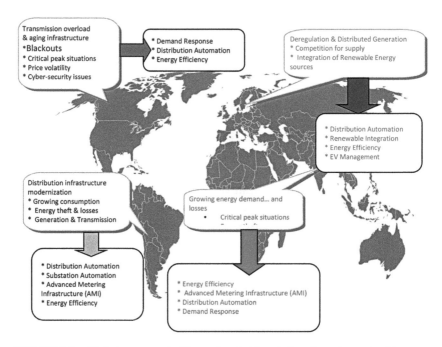

FIGURE 1.3   Global energy priorities and probable solution through smart grid.

1.6   ENABLING TECHNOLOGY AND SOPHISTICATED CONTROL

The recent technological advancement helps to connect and integrate the isolated technologies to achieve better and efficient energy management. The advanced communication system empowers the consumer and increases the participation of end user in economic dispatch of electricity along with main utility. Figures 1.4 and 1.5 depict a typical smart grid setup with complete connectivity to each and every source and load. The connectivity plays a vital role in making the grid smart by facilitative two-way energy flow. In a conventional grid, the energy flow is unidirectional, that is, from the

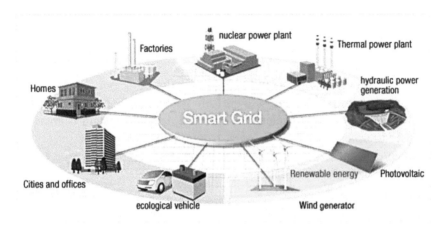

FIGURE 1.4   Smart grid connectivity.

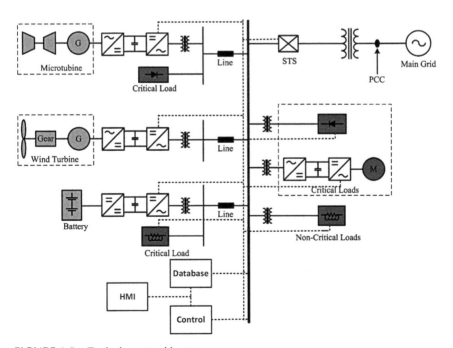

FIGURE 1.5   Typical smart grid setup.

utility to consumers. Whereas with the help of advanced technology, the end users are allowed to produce or store power through distributed generation and participate in the energy ecosystem. With the help of historical data and accurate forecasting mechanism, it is possible to implement efficient demand and response program by committing to reduce load when demand is high and allowing direct control of loads. The variable tariff programs can also be another aspect to encourage consumers to schedule their electricity consumption to avoid surge in demand during peak periods by increasing the tariff. Through the integration of smart building devices and systems, intelligent schemes can be used to perform automated load management to achieve desired energy efficiency. The cloud database system enables the end user to access their consumption pattern and energy pricing data; this will help consumer to predict energy needs, sell excess power, and isolate the sources of problems (Caamano et al., 2009).

1.7   CONCLUDING REMARKS

The growth in energy demand and recent advancement in technology are facilitating a smarter way to manage the energy ecosystem in this new era of smarter world. The paradigm shifts in electric power system technology changed the energy generation and consumption largely. Energy management systems, advanced control technology, energy storage, and intelligent information and communication platforms are enabling green, reliable, and economic energy supply. This new development maximizes financial benefits and reduction in environmental pollution by encouraging increased sustainable energy resources.

KEYWORDS

- **Internet of Things**
- **distributed energy sources**
- **demand and response management**
- **renewable energy generation**
- **energy storage**

REFERENCES

Caamano, E., et al. Interaction Between Photovoltaic Distributed Generation and Electricity Networks. *Progress Photovoltaic's Appl*. **2008,** *16*(7), 629–643.

Chowdhury, S.; Chowdhury, S. P.; Crossley, P. *Microgrids and Active Distribution Networks*; The Institution of Engineering and Technology: London, United Kingdom, 2009.

Etawil, M. A.; Zhao, Z. Grid-connected Photovoltaic Power Systems: Technical and Potential Problems—A Review. *Renew. Sustain. Energy Rev*. **2010,** *14*(1), 112–129.

Gaurav, S., et al. Energy Management of PV–Battery Based Microgrid System. *Procedia Technol*. **2015,** *21*, 103–111.

Kouro, S., et al. Recent Advances and Industrial Applications of Multilevel Converters. *IEEE Trans. Ind. Electron*. **2010,** *57*(8), 2553–2580.

Omran, W. A.; Kazerani, M.; Salama, M. M. A. In *A Study of the Impacts of Power Fluctuations Generated from Large PV Systems*, IEEE PES/IAS Conference on Sustainable Alternative Energy,Valancia, Spain, 2009, 1–6.

Such, M. C.; Cody, H. In *Battery Energy Storage and Wind Energy Integrated into the Smart Grid*, ISGT 2012, IEEE PES, Washington, United States, 2012, 1–4.

Teodorescu, R.; Liserre, M.; Rodrıguez, P. *Grid Converters for Photovoltaic and Power Systems,* 1st ed.; John Wiley: New Jersey, United States, 2011.

Trowler, D.; Bret, W. Bi-directional Inverter and Energy Storage System. *Texas Instrum. (Arkansas)* **2008,** 1–29.

CHAPTER 2

DESIGN OF PV SYSTEMS WITH BATTERY STORAGE

J. PRASANTH RAM and N. RAJASEKAR*

Solar Energy Research Cell (SERC), School of Electrical Engineering (SELECT), VIT University, Vellore 632014, Tamil Nadu, India

Corresponding author. E-mail: natarajanrajasekar@gmail.com

CONTENTS

ABSTRACT

Eco-friendly power generation from renewables has gained significant attention over the past decade. In fact, the government has laid a huge initiative by providing incentives to increase the power generation from renewables. Particularly, solar power generation from solar power parks are the trends of modern time. Since solar energy is a nonlinear current source that produce direct current , batteries play a vital role in saving the energy. In addition, the nonlinearity is solar characteristics need an maximum power point controller to extract efficient energy. To accomplish the aforementioned tasks, modeling, maximum power extraction, and effective energy management is required. Moreover, solar power can only be harvested during daytime hence, an effective energy management system is highly necessary to supply loads during day- and nighttime. Therefore, in this chapter, an overview about modeling, maximum power extraction, and an schematic explanation about a simple energy management system is explained in brief. Battery modeling and its characteristics is also been provided for clarity.

2.1 AN INTRODUCTION TO SOLAR ENERGY

Tremendous potential to supply large power demand, enormous availability, and continuous exhaustion of coal reserve have increased the penetration of renewable energy-based power generation. With recent smart grid technologies, electrification via renewable energy resources is perceived as one of the smartest way that could solve the energy crisis problem. In particular, remote area electrifications via solar photovoltaic (PV) systems are vital and gains more attention (Ram et al., 2017a).

Various ways by which power from solar energy is converted are solar thermal and solar PV systems. However, solar thermal power conversion requires additional cooling unit and external arrangements which will further increase the installation cost, whereas in case of solar PV, installions are simple and power conversion efficiency is far higher in comparison to the earlier. In addition, power generation via solar PVs is attractive due to the following reasons: (1) less maintenance, (2) zero pollution, (3) absence of moving and rotating parts, and (4) zero noise. Solar PV utilizes solar cell that converts incident photons energy to electricity. But on the other hand, PV suffers due to low efficiency of 17–20% (Ram et al., 2017b). However, multiple PV technologies indulging organic materials are under hopeful research to achieve higher efficiency. Since the PV characteristics and panel efficiency are very crucial in designing the PV system, the knowledge on usage of PV technologies will predominantly increase the efficiency of the system. Although PV systems are more resourceful and advantageous, the sustainability of the solar PV under lesser irradiation/absence of insolation during night time provoked the necessity to build an effective power management system (Liao et al., 2009). Thus, PV necessitates external/battery sources to store charges during effective sun hours and thereby supply to load during poor insolation conditions. This PV battery (PB) energy management system utilizes a power electronic (PE) interface to either connect to an alternating current (AC)/direct current (DC) load.

By utilizing various power converter topologies, several research in PV with battery storage systems are being put forward. Liao et al. (2009) have used DC–DC buck converter and a bidirectional converter to interface PV and battery. The proposed system has a common DC link where four modes of operation are successfully demonstrated. More importantly, the author operated PV in constant voltage and maximum power point tracking (MPPT) mode to effectively improve the power flow of the system. In another work, Mahmood et al. (2012) have proposed a different control strategy using a proportional integral (PI) controller, but the authors have used DC–DC boost converter in comparison to earlier work. In this work, an effective control over battery strength of charging (SOC) and voltage regulation at DC link are under limits to construct an effective energy management system. Considering the real-life load conditions, considerations, Kairies et al. (2016) have effectively proposed an average operating efficiency of a PB system under European operating conditions is estimated. Further, both DC and inverter connections with individual parameters are taken into the account to estimate the load profiles of the under farmland conditions.

Owing to the excellence in artificial intelligence techniques Magnor et al. (2016) have performed the optimization of battery sizing via genetic algorithms and compared their design with the conventional systems. The author has arrived specific boundaries and limits for battery operation with respect to SOC which has resulted an economic operation of the system. Similar work to increase the battery lifetime is being proposed by

Angenendt et al. (2016), for home storage systems. The author has considered different weather condition and analyzed the seasonal effects to the battery SOC. This system has proved its profitability in the PB system even under different irradiation conditions.

Irrespective of PV sources with battery applications, there are new hybrid topologies that are being the trends in autonomous microgrid environment. Further, these have an inherent advantage of utilizing power from any of the sources such as wind, solar, and fuel cell in failure of power generation from any of the earlier sources. Singh et al. (2016) have proposed and investigated the rural area and did a field study to experiment hybrid topology involving biomass, PV, and wind energy systems. In another approach, Ren et al. (2016) have operated an energy management having PV, wind, and battery under grid connected conditions, where it has a provision to operate in standalone mode as well.

In this chapter a brief explanation on solar PV modeling and its characteristics, battery modeling, importance of MPPT, and energy management with power converter topology are analyzed. Further, the necessity of MPPT operation in PV systems, battery operation, and different modes to operate PB hybrid system are also discussed in detail.

2.1.1   SOLAR PV MODELING

Despite the methodologies involved in fabricating the PV module, the modeling of solar PV has higher significance to deliver the optimal performance. Two common procedures followed in PV modeling are (i) analytical and (ii) numerical methods (Babu et al., 2015b). In analytical modeling, the nonlinear characteristics of solar PV are traced by considering the maximum power point in the I–V curve at different irradiations, whereas in numerical method, various instances of I–V curves are measured from the datasheet and matched with the obtained values. Among the two approaches followed, numerical method has higher accuracy because the data obtained are measured at each instance and over the I–V curve, and hence the exact reproduction of I–V curve is possible (Humada et al., 2016). But in the case of analytical method, improper selection of maximum power point leads to inaccurate I–V curve generation and, moreover, due to change in irradiation the operating point of PV varies continuously.

An accurate PV model should be highly sensitive to the change in irradiation and temperature levels, such that exact I–V characteristics can be replicated in a time-varying environment. However, the accurate reproduction of the I–V curve in agreement with manufacturer's data-sheet under all atmospheric conditions is a challenging task. To model a PV cell, data from manufacturers such as (i) open circuit voltage (V_{oc}), (ii) short circuit current (I_{sc}), (iii) maximum power voltage (V_{mp}), and (iv) maximum power current (I_{mp}) are mandatory. Making the scenario even worse, PV modeling further requires information regarding PV current (I_{pv}), diode saturation current (I_o), series resistance (R_s), parallel resistance (R_p), and diode ideality factor (a). Unfortunately, these data are not available from the manufacturer's data sheet and are unknown (Ram et al., 2017a). This problem is commonly referred to us as 'parameter identification problem,' and these unknown parameters are usually optimized using a suitable parameter extraction technique. Undoubtedly, to replicate exact PV characteristics through simulation, the necessity of an accurate PV model is highly imperative. Two common methodologies followed in PV cell modeling are single-diode modeling and double-diode modeling. Single-diode modeling is less complex and simple. Even though double-diode model is slightly complex with more parameters, it is highly accurate in reproducing exact PV characteristics. The schematic representation of single and double-diode model is represented in Figure 2.1a and b.

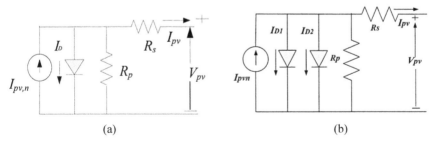

(a)   (b)

FIGURE 2.1   Schematic of single-diode model and double-diode model.

Applying KCL for a single-diode module and the output current equation of circuit is given by,

$$I_{PV} = N_{pp} \left\{ I_{PV,n} - I_O \left[\exp\left(\frac{V_{PV} + I_{pv}R_S}{V_t N_{ss}} \right) - 1 \right] \right\} - \left(\frac{V_{PV} + I_{pv}R_S}{R_P} \right), \qquad (2.1)$$

where '$I_{PV,n}$' is the PV panel current, 'I_O' is the reverse saturation current, 'R_S and R_P' are the series and parallel resistance, and 'V_t' is the thermal voltage at any temperature. 'N_{ss}' and 'N_{pp}' are the number of cells connected in series and parallel to form a module. In single-diode modeling, five parameters are unknown ($I_{PV,n}$, I_D, R_S, R_P, and a). The output equation for a double-diode model is given in the following equation:

$$I_{PV} = N_{pp} \left\{ I_{PV} - I_{01} \left[\exp\left(\frac{V_{PV} + IR_S}{a_1 V_t N_{ss}} \right) - 1 \right] - I_{02} \left[\exp\left(\frac{V_{PV} + IR_S}{a_2 V_t N_{ss}} \right) - 1 \right] \right\}$$
$$- \left(\frac{V_{PV} + I_{PV}R_S}{R_P} \right), \qquad (2.2)$$

where I_{01} and I_{02} are the reverse saturations and 'a_1 and a_2' are the diode ideality constants for diode 1 and 2. This parameter in single diode and double-diode model decides the solar PV effectiveness. In double-diode model, seven parameters are unknown (a_1 and a_2).

From the above discussion, it is clear that modeling of solar PV involves accurate two-diode model and a simple single-diode model. In both the models, 'Ipv, n' and reverse saturation current ('Io') in most of the cases are calculated. Also, analytical calculation of 'Ipv, n' and 'Io' avoids the computational burden. The formula used for PV current calculation and finding reverse saturation current is presented in eqs 2.4–2.6.

$$I_{PV,n} = (I_{scn} + K_i \Delta T) \frac{G}{G_n}, \qquad (2.3)$$

where 'I_{scn}' is the nominal short circuit current, 'K_i' is the temperature current coefficient, and 'G_n' is the nominal irradiance at standard test conditions (STC)—(1000 W/m²) and $\Delta T = T - T_n$, where 'T_n' is the nominal temperature at STC (25°C). Generally, the nominal values such as I_{scn}, V_{ocn}, K_i, K_v, and P_{max}, will be given in the manufacturer's data sheet. The reverse saturation current for the single-diode and double-diode model can be calculated as follows:

for single-diode model,

$$I_o = \frac{I_{PV}}{\exp[(V_{oc} + K_v \Delta T) / a / V_T] - 1} \tag{2.4}$$

for double-diode model,

$$I_{o1} = \frac{I_{PV}}{\exp[(V_{oc} + K_v \Delta T) / a_1 / V_T] - 1} \tag{2.5}$$

$$I_{o2} = \frac{I_{PV}}{\exp[(V_{oc} + K_v \Delta T) / a_2 / V_T] - 1}, \tag{2.6}$$

where 'a' is the diode ideality factor for a single-diode model while 'a_1 and a_2' are the ideality factors in double-diode model, and 'K_v' is the short circuit voltage ratio.

2.1.2   SIGNIFICANCE OF PV PARAMETERS ACCURACY

In order to show the effect of series, parallel resistance, and ideality factor over the nonlinear I–V curve, different values of 'Rs, Rp, and a' values are considered for plotting the I–V curve. PV characteristics represented in Figure 2.2a–c corresponding to different values have following observations: (i) 'Rs' value has to be lower to reproduce accurate I–V characteristics, (ii) 'Rp' has been maximized to improve the higher fill factor and efficiency (Ismail et al., 2013), and (iii) increase in 'a' leads to dragged I–V curve generation. In general, the 'Rs' represents the sum of resistance which combines the metallic contacts, semiconductor leads, and interconnections of PV cells in the module and these drops are just an I^2R drop. Hence, Rs is always kept minimum and similarly 'Rp' is caused due to recombination losses which improves the conductivity of semiconductor hence 'Rp' is always kept maximum (Babu et al., 2016a).

Scholarly research in modeling of solar PV has computed their parameters via optimization procedures. Several approaches that have been put forward to calculate the PV parameters are analyzed and compared in Table 2.1.

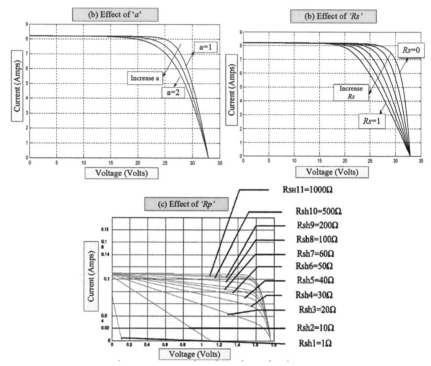

FIGURE 2.2   (a) Effect of '*Rs*,' (b) '*Rp*,' and (c) '*a*.'

2.1.3   *SOLAR PV MODULE AND ITS CHARACTERISTICS*

In general, PV module is highly dependent to the irradiation, where the current flow in the PV module is controlled. Hence, it is called as a current controlled device. Similarly, voltage across the PV module is greatly influenced by temperature variation. To experiment the earlier effect of change in current under insolation change, SM55 multicrystalline PV module is considered and its characteristics for different irradiation are shown in Figure 2.3a. In continuation, the effect of insolation change affecting the power output is also shown in P–V characteristics. For understanding, the datasheet of SM55 module is shown in Table 2.2. Unlike insolation change, temperature changes in solar PV may not experience a huge change in power/voltage but, I–V and P–V characteristics in Figure 2.3b have reasonable power and voltage difference, thus, illustrating both irradiation and temperature affect power production.

TABLE 2.1 Summary of Optimization Techniques used to Calculate Modeling Parameters.

Sr. no.	Author name	Optimization method used	Optimized parameters	Single-diode model/double-diode model	Method of computation	PV modules modeled
1	Ismail et al. (2013)	Genetic algorithm and interior point method	$Ipvn, Io/(Io1$ and $Io2), Rs, Rp,$ and a	Both single and double-diode model	Analytical	Kyocera Kc200GT and ST40
2	Askarzadeh et al. (2013b)	Bird mating optimizer	$Ipvn, Io/(Io1$ and $Io2), Rs, Rp,$ and a	Both single and double-diode model	Numerical	RTC France 70 mm PV cell
3	Askarzadeh et al. (2012a)	Harmony search	$Ipvn, Io/(Io1$ and $Io2), Rs, Rp,$ and a	Both single and double-diode model	Numerical	RTC France 70 mm PV cell
4	Askarzadeh et al. (2013c)	Artificial bee swarm optimization	$Ipvn, Io, Rs, Rp,$ and a	Single-diode model	Numerical	RTC France 70 mm PV cell
5	AlHajri et al. (2012)	Pattern search	$Ipvn, Io/(Io1$ and $Io2), Rs, Rp,$ and a	Both single and double-diode model	Numerical	RTC France 70 mm PV cell and module
6	Nassareddine et al. (2016)	Lambert W function	$Rs, Rp,$ and a	Single-diode model	Numerical	RTC France PV cell, SQ80, ST40, and Kc200GT
7	Alam et al. (2015)	Flower pollination algorithm	$Rs, Rp, a/a1$ and $a2$	Both Single and double-diode model	Numerical	RTC France PV cell, SM55, ST40, and KC200GT
8	Babu et al. (2015b)	Fireworks algorithm	$Rs, Rp, a1$ and $a2$	Double-diode model	Analytical	SP70, SM55, and KC200GT
9	Rajasekar et al. (2013)	Bacterial foraging algorithm	$Rs, Rp,$ and a	Single-diode model	Analytical	SP70, SM55, and S36
10	Ishaque et al. (2011)	Differntial evolution	$Rs, Rp,$ and a	Single-diode model	Analytical	S75, SM55, and ST40

TABLE 2.2 Datasheet of SM55 Module.

Sr. no.	Panel details		
1	Maximum power rating P_{max}	[Wp]	55
2	Rated current I_{MPP}	[A]	3.15
3	Rated voltage V_{MPP}	[V]	17.41
4	Short circuit current I_{SC}	[A]	3.45
5	Open circuit voltage V_{OC}	[V]	21.7

2.1.4 IMPORTANCE OF LOCATING MAXIMUM POWER POINT IN SOLAR PV

From the I–V characteristics modeled in Figure 2.3, it can be observed that the PV characteristic is nonlinear. Since the energy cost per watt and the installation costs are higher for PV power plants, it is highly essential to extract the maximum available power from the system. From Figure 2.3a and b, it is clear that there exists an unique operating point [marked as maximum power point (MPP)] in I–V and P–V curves which keeps on shifting with respect to the irradiation and temperature levels. This point is the maximum power point at which the maximum power can be extracted from the panel. It is highly necessary to make the panel operate at this point with respect to the change in irradiation and temperature levels. Any improper operation may lead to reduced PV efficiency. Thus, MPPT controllers form an integral part of the PV system by tracking the MPPs at varying environmental conditions to ensure maximum power transfer. As shown in Figure 2.4, one of the best ways to implement MPPT controller is by introducing a PE interface

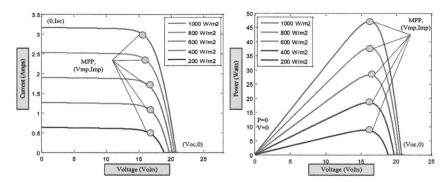

FIGURE 2.3a I–V and P–V characteristics of 55 W panel for different irradiations.

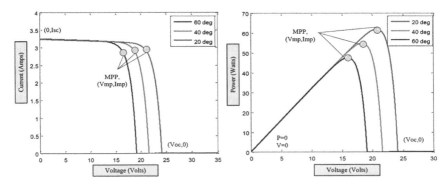

FIGURE 2.3b   I–V and P–V characteristics of 55 W PV panel for different temperatures.

between PV source and load. The presence of the controller effectively alters the resistance seen by the panel by changing the duty cycle of the converter with respect to the current and voltage sensed from the panel and hence impetus the panel to operate closer to MPP.

For a clear understanding, the implementation of MPPT controller for the stand-alone application with power converter is shown in Figure 2.4. In order to explain the importance of MPP operation, the I–V characteristic plotted for 1000 and 800 W/m² is utilized. To demonstrate the function

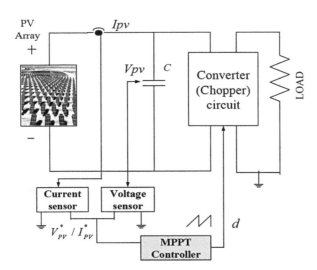

FIGURE 2.4   DC–DC converter with MPPT controller.

of MPPT controller both the I–V curve and load line characteristics are plotted in Figure 2.5. Let us assume that a PV module is directly coupled to a resistive load, the operating point of the load in the I–V curve is the intersection of the line with the I–V curve. The slope of the load line (1/R) is also represented. It can be seen that as the slope of the load line varies, the operating point shifts either way. For instance, operating point shifts from point 'a' to point 'b' when the load resistance changes from 6.35 to 2 Ω. Likewise, operating point moves from point 'c' to 'b' when irradiation changes from 1000 to 800 W/m².

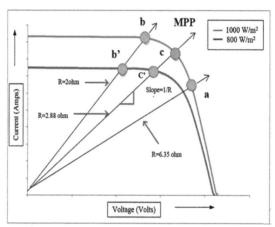

FIGURE 2.5 I–V characteristics of a PV panel connected to different load conditions.

Hence, it can be summarized that by either changing the load resistance or by changing the I–V characteristics of the panel, the operating point can be altered. Unfortunately, in real-time conditions, neither the load connected nor the solar PV characteristics can be altered. Hence, it is mandatory to introduce a PV interface which can effectively alter the effective resistance seen by the PV panel. Hence, a power converter is introduced between load and solar PV. By varying the duty cycle of the converter with respect to the sensed voltage and the current from the panel, the input resistance of the converter seen by the panel can be varied to extract the maximum power available at every instant. Hence, every PV system has a PE interface connected between the PV and the load which guarantees PV panel operation at higher efficiencies.

2.1.5 PV UNDER PARTIAL SHADED CONDITIONS

Generally, PV array is formed by connecting panels in series and parallel fashion to meet the energy demand. This interconnection between panels vary according to the type of configuration (e.g., series parallel, bridged link, and total cross ties), but the common phenomenon that affects the PV array output is partial shading. Partial shading conditions (PSC) are the phenomenon that occurs in a PV array due to the passage of cloud, tree, and building shadows.

Occurrences of partial shading create hotspots on the lesser irradiated panel, thereby, creating voltage drops across the panel's output. With an intention to protect PV panels in PV array, every module is connected to a bypass diode connected in parallel and a blocking diode in series to avoid hotspot and current reversal problems, respectively. For illustration, the occurrence of partial shading and its consequential effect, a series of 4S PV panel connected in series string is constructed as shown in Figure 2.6 and its P–V and I–V characteristics are plotted in Figure 2.7a and b. The PV array is shown for two different environmental conditions: (i) uniform

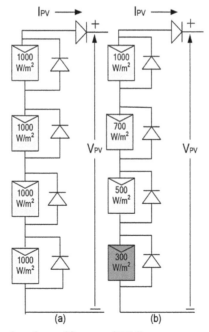

FIGURE 2.6   PV panel under uniform and PSC.

irradiation and (ii) PSC (Sangeetha et al., 2016a; Sangeetha et al., 2016b). Having two different atmospheric conditions, that is, patterns (a) and (b), the nature of I–V and P–V characteristics vary. Especially under the PSC, multiple peaks in the P–V curve occur and steps are created in I–V curves. While in case of uniform irradiation conditions single peak occur. The occurrence of multiple peak due to PSC might attribute to reduce power output, if the MPPT is not implemented in the PV system (Babu et al., 2016c).

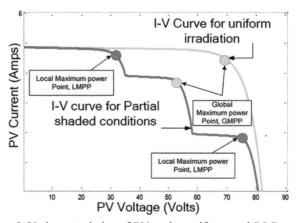

FIGURE 2.7a   I–V characteristics of PV under uniform and PSC.

under uniform and PSC.

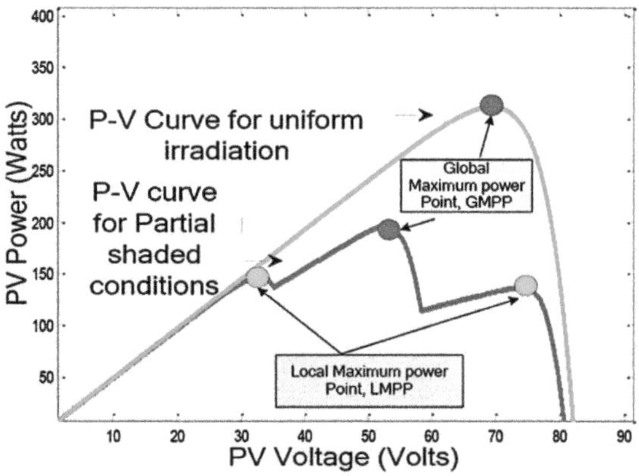

FIGURE 2.7b   P–V characteristics of PV under uniform and PSC.

2.2   BATTERY MODELING

2.2.1   SIMPLE AND IMPROVED BATTERY MODEL

Modeling of batteries is really crucial and it is witnessing a stage by stage development over a period of years. A simple way of modeling battery is by representing the voltage source (Eo) in series invariable resistance. The schematic of the simple battery model is represented in Figure 2.8a. where the open circuit voltage is represented as 'V_{oc}' and the output voltage 'V_b' can be obtained across the resistance 'R_b.' Since the battery model does not incorporate any internal resistance which has a higher effect to SOC, it cannot be applicable to all the applications but is preferred where SOC is not given priority. Taking SOC into account, the improvised model by replacing the constant 'R_b' with variable 'R_b' is being proposed. The equation used to model 'R_b' in the improvised model is given in (Dürr et al., 2006) the following.

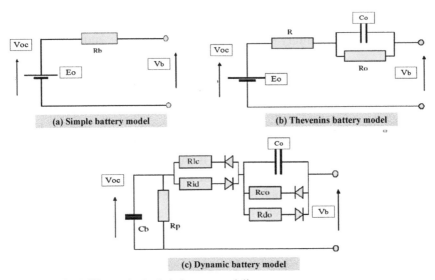

(a) Simple battery model

(b) Thevenins battery model

(c) Dynamic battery model

FIGURE 2.8   Different methods in battery modeling.

$$R_b = \frac{R_o}{S^k},$$ (2.7)

where
$$S = 1 - \frac{Ah}{C_{10}}, \tag{2.8}$$

where 'C_{10}' is the capacity for 10 h at reference temperature, 'A' is the discharging current, and 'h' is the time for discharging. In eq 2.7, 'R_o' is the fully charged battery resistance, 'S' is the state of the charge factor which varies from 0 to 1 depending on zero fully charged mode, and 'k' is the capacity coefficient.

2.2.2   THEVENINS BATTERY MODEL

This model is just a simple model which comprises additional 'Co' and 'Ro' to the equivalent circuit, where the 'Co' is included to represent the parasitic capacitance present in the battery and 'Ro' represents the resistance of the plate contacts and the electrolyte. Alike simple battery model, Thevenins also has the problem of fixed values where the SOC and discharge rate cannot be realized. However, in reality, these factors are mandatory under varying conditions. Due to the limitations, the model has less recognition in battery electrochemistry (Dürr et al., 2006).

2.2.3   DYNAMIC BATTERY MODELING

To carry out a practical discharging and charging of the battery, a realistic dynamic modeling with different design considerations involving self-charge and over-charge resistance is characterized as shown in Figure 2.8c. Keeping the SOC as objective, the elements of battery are modeled in the function of 'Voc' which is in relation to SOC (Dürr et al., 2006). The descriptions and the function of the different elements used in the dynamic battery modeling are presented in Table 2.3.

The dynamic model parameters are modeled by considering individual variations with respect to Voc. This ensures the functions of elements are in relation with Voc which is a relation of SOC. Hence, the model becomes more accurate. The capacitance determined based on Voc is given in eq 2.9.

$$C = k^* e^{\,(W*(Vm-Voc))F}, \tag{2.9}$$

where 'C' is the capacitance in Farad, 'k' is the gain factor, 'W' is the width factor, 'V_m' is the mean voltage, 'Voc' is the open circuit voltage, and 'F' is the flatness factor.

TABLE 2.3   Description of Elements used in Dynamic Battery Modeling.

Element	Name	Descriptions
C_b	Battery capacitance	It is modeled as a control voltage source in relation to SOC
R_p	Self-discharge resistance	It is modeled for accounting small leakage current in function of Voc
R_{ic} and R_{id}	Internal resistance during electrolyte and battery plates	The resistances are modeled to measure the drops at battery plate and electrolyte where it varies for charging and discharging
R_{oc} and R_{od}	Voltage drop resistance during charging and discharging	To measure the overvoltage drop during charge and discharge
C_o	Double-layer capacitance	Represents the behavior of battery under both charging and discharging

2.2.4 BATTERY CHARACTERISTICS

To explain the battery characteristics, an ideal 12 V battery at full charge mode is considered and its (a) capacity retention characteristics at fully charged mode and (b) total voltage (V/s) discharge time characteristics is plotted as shown in Figure 2.9a and b. From the retention characteristics presented in Figure 2.9a, it can be understood that fully charged battery has the ability to store charges for months; however, the SOC of the battery is steadily reduced in proportional to increase the number of months. Notably, it is important to mention that SOC highly depends on room temperature and the battery discharge is faster under high temperature is an important inference from the capacity retention characteristics. Generally, a battery under fully charged mode can be identified when the voltage across its terminals are hyped over the actual datasheet value. For example, a 12 V battery at the fully charged mode having 12.8 V as Voc is found and subsequently the reduced voltage under discharging mode at regular intervals is considered for study as shown in Figure 2.9b. From the terminal voltage (V/s) discharge time characteristics plotted for 25°C, it is observed that reduced Voc has faster discharging time compared to the fully charged *Voc*. Thus, it can be found that battery having high SOC discharge at slower rate where the *Voc* is notably high. On the other hand, battery with lower SOC has vice versa characteristics of the earlier.

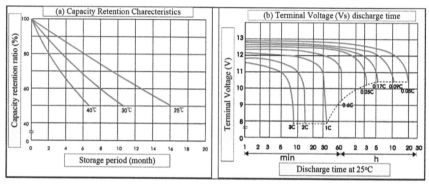

FIGURE 2.9 (a) Capacity retention characteristics and (b) terminal voltage (V/s) discharge time characteristics.

2.3 PV WITH BATTERY STORAGE SYSTEMS

The PV systems, though being sustainable, are highly dependent on climatic changes, and hence PV source may not able to satisfy the load requirement. Moreover, the absence of PV during less irradiated conditions will be a major concern to meet the required load demand. Hence, PV system demands for an effective energy management system by utilizing batteries and a PE interface (Liao et al., 2009). Various configurations utilized to construct an effective PV system design with battery storage are discussed here.

In general, for power systems, maintaining DC link voltage is crucial and the renewable sources on the supply side highly subject to environmental changes, thus PV requires a supplementary power source such as batteries to effectively construct a power management system. Considering a stand-alone PV system, a PB-based energy management strategy is constructed in Figure 2.10. The system consists of DC–DC buck converter and a bidirectional converter (Mahmood et al., 2012). The bidirectional converter operates in either buck or boost mode depending on the supply availability. In a conventional power system, battery can be directly connected to the DC bus. But, taking charge and discharging of the battery into the account, the bidirectional converter is connected on the load side. If a PV can deliver more power in excess to the requirement, the bidirectional converter operates in a buck mode to charge the battery and the same converter operates in boost mode, if PV power is insufficient to satisfy the load. The PB management system has the following advantages: (i) bidirectional converter has the complete control over battery charging and draining, (ii) the system is simple and easy for implementation, and (iii) an efficient power management strategy is put forward by utilizing unidirectional DC–DC and bidirectional converter.

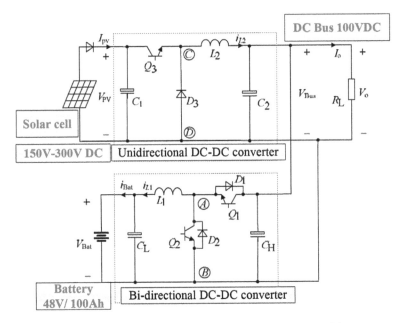

FIGURE 2.10 Power management control strategy using PV and battery.

2.3.1 OPERATING MODES OF PB SYSTEM

The PV system should maintain DC bus voltage of 100 V, hence the DC–DC converter which can operate at a wide range between 150 and 300 V is chosen. The battery is chosen to supply 48 V charged operating conditions and the whole system is designed to supply either DC or AC load except the fact that an inverter is to be connected in case of an AC system. On the basis of the limits of battery, DC link voltage, minimum and maximum operating charging conditions, there exist four operating modes for the configuration shown in Figure 2.10. To explain the operating modes of the PB system, the following specifications are defined. Also, it is important to define two modes of operation that happen to PV possible (a) MPPT mode and (b) constant voltage mode. During earlier, PB operates to charge the battery and supplies to load at the same time and in the later PV just operates to maintain DC link voltage.

V_{PV} = operating voltage of a PV array

V_{BAT} = operating voltage of the battery

$V_{BAT,\,max}$ = maximum battery voltage is set to be at 56 V

$V_{BAT,\,min}$ = minimum battery voltage is set to be at 44 V

I_{BAT} = charging and discharging of the battery

$I_{BAT} > 0$ = charging of battery, $I_{BAT} < 0$ = discharging of battery

$V_{VP,\,min}$ = minimum operating voltage of PV is set to be 150 V

$V_{BAT,\,max}$ = maximum operating voltage of PV is set to be 300 V

Mode (a): In this mode two possible ways of operation are possible, (1) PV source is greater than the output power ($P_O < P_{PV}$), that is, both the DC–DC converter and the bidirectional converter operates in buck mode. In this mode, solar PV operates at MPP so that it can supply to load and battery, where charging and discharging happens simultaneously. For understanding, the charging and discharging of the battery is represented by arrows as shown in Figure 2.11a. In mode (2) PV source is less than output power ($P_O < P_{PV}$), that is, the bidirectional converter operates in boost mode to match the load requirement.

Mode (b): In this mode battery reaches overcharged at $V_{BAT,\,max}$ and I_{BAT} reaches to 10 A in mode (a), hence the bidirectional converter switches to boost mode and the PV now operates in constant voltage mode. In this mode battery discharges and PV supplies to satisfy DC link voltage as shown in Figure 2.11b.

Mode (c): The mode is applicable when the PV source is not available, that is, this condition is experienced during dark light or nighttime. During this period, the battery supplies regularly to maintain 100 V in DC bus thereby, operating bidirectional converter in boost mode. The schematic of this mode of operation is shown in Figure 2.11c.

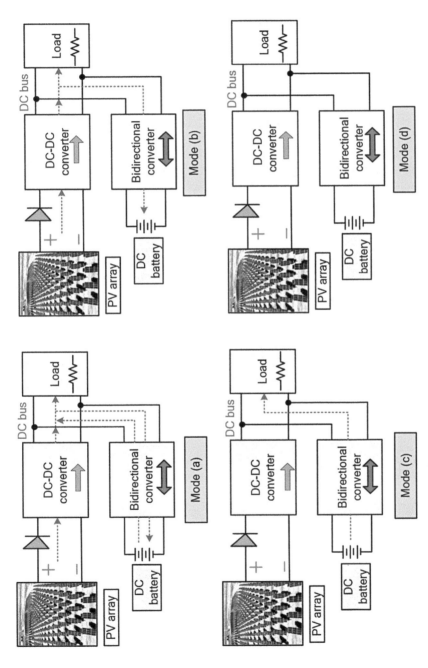

FIGURE 2.11 Power flow topologies in PV battery system (Mahmood et al., 2012).

Mode (d): Due to continued poor light and insufficient insolation, the battery may continuously operate at mode (c) and reaches to $V_{BAT,\,max}$, that is, battery voltage less than 44 V. Thus, the bidirectional converter should stop operating and the entire system is shut down. A similar occurrence is also possible in mode (a), when battery continues to operate under poor insolation, the bidirectional converter should stop operating at the same similar to the mode (d). The representation of PB hybrid system operated in this mode is given in Figure 2.11d.

2.3.2 ENHANCED CONTROL TECHNIQUE FOR PB SYSTEM

The control scheme is an extension of the conventional PV design, which is discussed earlier. Here, a DC–DC boost converter is used as unidirectional controller and two DC link capacitance are provided as one directly connected to the battery and other is connected to PV source via DC–DC boost converter. The schematic representation of the enhanced PV control

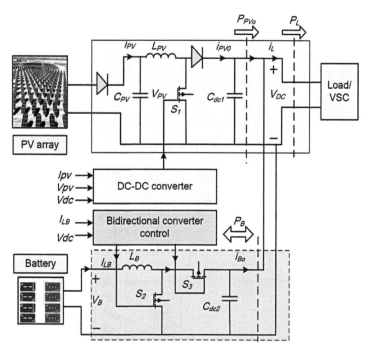

FIGURE 2.12   Enhanced control strategy for PV battery hybrid system (Liao et al., 2009).

strategy is shown in Figure 2.12. Three operating modes of this configuration are (a) normal, (b) PV/SOC regulation, and (c) PV/DC link regulation. The advantages of this system are (i) closed loop control, (ii) effective to system changes, and (iii) flexible to maintain SOC of the battery during charging and discharging.

2.4   DESIGN OF PB HYBRID SYSTEM

Problem 1:

Design a solar power and water pump system for a landowner having 216 sheep on 40 acres of pasture land where a sheep consumes 5 gal/day. The intake line is located at 42 ft below the base of the tank. The landowner intends to gravity feed two watering troughs located 1118 and 712 ft from the proposed storage tank. Two 500 gal troughs are to be used. To store the water, a storage tank is being used having 8′ height from the ground level where it can store 3 days of required water. Design the suitable PV array structure of the pump pipeline shown in Figure 2.13.

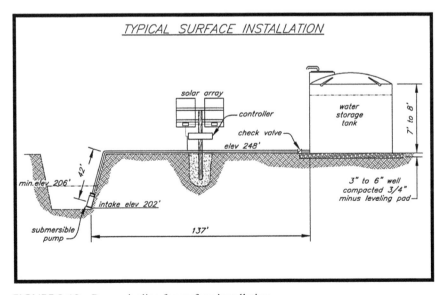

FIGURE 2.13   Pump pipeline for surface installation.

Solution:

Step 1: water requirement: the water requirement for the sheep in the pasture land can be calculated as follows:

$$216 \ sheep \times 5 \ gal/day/animal = 1080 \ gal/day$$

Step 2: water Storage

Since, the system's total water storage capacity should be sufficient for a minimum of 3 days water use, the minimum storage capacity calculated in step 1 is used.

$$1080 \ gal/day*3 \ days = 3240 \ gal$$

Two 500 gal water troughs are included in the system, providing a total storage capacity of 1000 gal (2 × 500 = 1000). Therefore, the storage tank must be sized to hold a minimum of 2240 gal (3240 − 1000 = 2240). On the basis of information from different distributors, a 2500 gal water tank is a readily available size. The tank is 92 in. in diameter and 95 in. tall. As a safety precaution, it is recommended that the tank and troughs be filled prior to use to ensure that the system has adequate water storage.

Step 3: solar insolation and PV panel location

The solar insolation values for the respective are arrived based on the proximity of the two locations. Therefore, the solar insolation values for an approximate of 5 months will be used for this design as shown below. These values are useful in arriving effective sun hours for the process.

$$Effective \ Sun \ Hours = \frac{(5.6 + 6.0 + 6.7 + 6.3 + 5.4)}{5} = 6.0 \, h$$

Step 6: design flow rate for the pump

The design flow rate of the pump can be measured in terms of *gallons per minute (gpm)* by utilizing the data as shown below:

$$Gallons \ per \ minute = \frac{1080 \ gal}{6.0 \, peak \, h * 60 \, min/ \, h} = 3.0 \, gpm$$

Step 7: total dynamic head (TDH) for the pump

The TDH of the pump can be founded by using the formulae displayed below:

$$TDH = Vertical\ Lift + Pressure\ Head + Friction\ Losses$$

Vertical lift is the vertical distance between the water surface at the intake point (the stream's water surface) and the water surface at the delivery point (the tank's water surface). In this example:

$$Vertical\ Lift = 248\ ft + 8\ ft - 206\ ft = 50\ ft$$

Pressure head is the pressure at the delivery point in the tank. Hence, there is no pressure at the delivery point so:

$$Pressure\ Head = 0\ ft$$

Friction loss: The total friction losses in the pipeline are minimal. As such, a less expensive, smaller diameter pipe is selected. Approximately 137 ft of ¾ in. diameter PVC pipe will be used to convey water from the source to the tank. The friction loss for ¾ in. pipe conveying 3.0 gpm (Step 6) is approximately 2.17 ft of head loss per 100 ft of pipe. Therefore, the total estimated friction loss for 137 ft of pipe is 2.97 ft (137 ft ÷ 100 ft of pipe × 2.17 ft of head loss/100 ft = 2.97 ft). Minor losses through elbows and valves are estimated to be 1.8 ft for a total friction loss of 4.77 ft (2.97 + 1.8 = 4.77).

$$TDH = 50\ ft + 0\ ft + 4.77\ ft = 54.77\ ft\ approx.\ use\ 55\ ft$$

Step 8: pump selection and associated power requirement

The pump can be selected by comparing the design flow rate and TDH calculated in Steps 6 and 7 with the information from the manufacturer's pump curves. The first step for this example is to locate the design flow rate of 3.0 gpm on the y axis of the pump curve diagram and draw a horizontal line across the chart through this point, as shown. Next, locate where this line intersects the curve representing a TDH of 55 ft (60 ft being the closest curve in the case of Figure 2.14). From this point of intersection, draw a vertical line to the bottom of the graph. The point where the vertical line crosses the x axis shows the peak power requirement for the pump. As shown, based on a calculated flow rate of 3.0 gpm and a TDH of 55 ft (rounded up to 60 ft), a minimum input of 160 W of peak power is required.

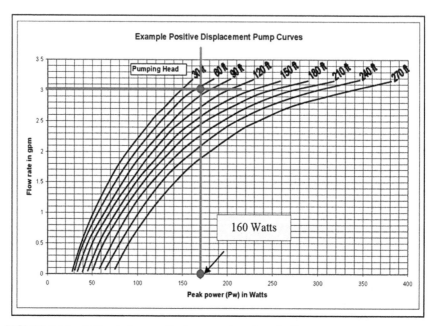

FIGURE 2.14 Estimated power for the pump based on the flow rate.

Step 9: *PV design*: power required (performance curve): 160 W

Assumed solar panel rating: *234 W, 71 V, and 3.3 A*
 The panel information for the system can be described as:

Number of Panels: *2 wired in series*

Problem 2

Design a hybrid solar PV system to supply the following loads:

No. of CFLs: 4 (30 W)	No. of running hours: 8 h
No. of DC fans: 2 (45 W)	No. of running hours: 10 h
No. of TV set: 1 (200 W)	No. of running hours: 4 h
No. of refrigerator: 1 (500 W)	No. of running hours: 16 h

*note: CFL (Compact Fluorescent Light)

Determine the size of the PV array to supply the above loads assuming days of autonomy and sun peak hours as 2 and 5.6, respectively. All appliances have nominal voltage of 24 V.

Solution

Sr. no	Element	Qty	W	No of running h	W-h
1	CFL	4	30 W	4	960
2	DC Fans	2	45 W	10	900
3	Television	1	200 W	4	800
4	Refrigerator	1	500 W	16	8000
Total Watts			775 W	Total W-h	10,660

Total watts required = 775 W

Total W-h required = 10,660

Step 1: Estimation of current

Required current = Total W/system nominal voltage = 775/24 = 5.6 A

Step 2: Estimation of number of A-h:

A-h = Total W-h/system nominal voltage = (10,660/24) = 444.6 W-h

Step 3: Battery sizing

$$Total\ number\ of\ A\text{-}h\ needed\ for\ battery =$$
$$(Estimated\ A\text{-}h*Estimated\ Sun\ hours)/Days\ of\ Autonomy =$$
$$(444.6*2)/0.6 = 1480.5\ A.h$$

Considering safety factor and 90% efficiency

$$Total\ W\text{-}h/0.9 = 1480.5/0.9 = 1645.03\ A\text{-}h$$

Considering the battery of 12 V *Voc* and 50 A-h capacity

No of battery connected in series = 24/12 = 2

No of batteries connected in parallel = 1645.03/50 = 11

Step 4: Estimating PV sizing

Assuming the following data: *Pmp = 55W, Vmp = 17.4V, and Imp = 3.3 A.*

$$No \ of \ PV \ panels \ to \ be \ connected \ in \ series = \frac{24}{17.4 = 1.3 = 2 \ (approx)}$$

$$No \ of \ batteries \ connected \ in \ parallel = \frac{5.41}{3.3 = 1.61 = 2 \ (approx)}$$

2.5 CONCLUSION

In this chapter, the design of PV systems with battery storage is presented. Battery modeling along with PV modeling is explained in brief to get the basic understanding of PV and battery. Different configurations and the necessity of PE to connect between PV and battery is exclusively presented with different operating modes. Also, the necessity of PV to be operated in MPP under PSC using a load line characteristics is demonstrated with its characteristics.

KEYWORDS

- **photovoltaic (PV)**
- **renewable energy resources**
- **partial shading condition**
- **maximum power point**
- **global maximum power point**

REFERENCES

Alam, D.; Yousri, D.; Eteiba, M. Flower Pollination Algorithm Based Solar PV Parameter Estimation. *Energy Convers. Manag.* **2015**, *101*, 410–422.

AlHajri, M.; El-Naggar, K.; AlRashidi, M.; Al-Othman, A. Optimal Extraction of Solar Cell Parameters Using Pattern Search. *Renew. Energy* **2012**, *44*, 238–245.

Angenendt, G.; Zurmühlen, S.; Mir-Montazeri, R.; Magnor, D.; Sauer, D. U. Enhancing Battery Lifetime in PV Battery Home Storage System Using Forecast Based Operating Strategies. *Energy Procedia* **2016**, *99*, 80–88.

Askarzadeh, A.; Rezazadeh, A. Parameter Identification for Solar Cell Models Using Harmony Search-based Algorithms. *Sol. Energy* **2012a**, *86*, 3241–3249.

Askarzadeh, A.; Rezazadeh, A. Extraction of Maximum Power Point in Solar Cells Using Bird Mating Optimizer-based Parameters Identification Approach. *Sol. Energy* **2013b,** *90,* 123–133.

Askarzadeh, A.; Rezazadeh, A. Artificial Bee Swarm Optimization Algorithm for Parameters Identification of Solar Cell Models. *Appl. Energy* **2013c,** *102,* 943–949.

Babu, T. S.; Rajasekar, N.; Sangeetha, K. Modified Particle Swarm Optimization Technique Based Maximum Power Point Tracking for Uniform and Under Partial Shading Condition. *Appl. Soft Comput.* **2015b,** *34,* 613–624.

Babu, T. S.; Ram, J. P.; Sangeetha, K.; Laudani, A.; Rajasekar, N. Parameter Extraction of Two Diode Solar PV Model Using Fireworks Algorithm. *Sol. Energy* **2016a,** *140,* 265–276.

Babu, T. S.; Sangeetha, K.; Rajasekar, N. Voltage Band Based Improved Particle Swarm Optimization Technique for Maximum Power Point Tracking in Solar Photovoltaic System. *J. Renew. Sustain. Energy* **2016c,** *8*(1), 013106.

Dürr, M.; Cruden, A.; Gair, S.; McDonald, J. R. Dynamic Model of a Lead Acid Battery for Use in a Domestic Fuel Cell System. *J. Power Sources* **2006,** *161*(2), 1400–1411.

El-Naggar, K.; AlRashidi, M.; AlHajri, M.; Al-Othman, A. Simulated Annealing Algorithm for Photovoltaic Parameters Identification. *Sol. Energy* **2012,** *86,* 266–274.

Humada, A. M.; Hojabri, M.; Mekhilef, S.; Hamada, H. M. Solar Cell Parameters Extraction Based on Single and Double-diode Models: A Review. *Renew. Sustain. Energy Rev.* **2016,** *56,* 494–509.

Ishaque, K.; Salam, Z. A Comprehensive MATLAB Simulink PV System Simulator with Partial Shading Capability Based on Two-diode Model. *Sol. Energy* **2011,** *85,* 2217–2227.

Ismail, M.; Moghavvemi, M.; Mahlia, T. Characterization of PV Panel and Global Optimization of Its Model Parameters Using Genetic Algorithm. *Energy Convers. Manag.* **2013,** *73,* 10–25.

Kairies, K. P.; Haberschusz, D.; Wessels, O.; Strebel, J.; van Ouwerkerk, J.; Magnor, D.; Sauer, D. U. Real-life Load Profiles of PV Battery Systems from Field Measurements. *Energy Procedia* **2016,** *99,* 401–410.

Khoury, J.; Mbayed, R.; Salloum, G.; Monmasson, E. Design and Implementation of a Real Time Demand Side Management Under Intermittent Primary Energy Source Conditions with a PV-battery Backup System. *Energy Build.* **2016,** *133,* 122–130.

Liao, Z.; Ruan, X. In *A Novel Power Management Control Strategy for Stand-alone Photovoltaic Power System,* Power Electronics and Motion Control Conference 2009 (IPEMC'09), IEEE 6th International, May 17, Wuhan, China, 2009; pp 445–449.

Magnor, D.; Sauer, D. U. Optimization of PV Battery Systems Using Genetic Algorithms. *Energy Procedia* **2016,** *99,* 332–340.

Mahmood, H.; Michaelson, D.; Jiang, J. In *Control Strategy for a Standalone PV/Battery Hybrid System,* IECON 2012 38th Annual Conference on IEEE Industrial Electronics Society, 2012; pp 3412–3418 (IEEE).

Nassar-Eddine, I., A. Obbadi, Y. Errami, and M. Agunaou. "Parameter estimation of photovoltaic modules using iterative method and the Lambert W function: A comparative study." Energy Conversion and Management 119 (2016): 37-48.Rajasekar, N.; Kumar, N. K.; Venugopalan, R. Bacterial Foraging Algorithm Based Solar PV Parameter Estimation. *Sol. Energy* **2013,** *97,* 255–265.

Ram, J. P.; Rajasekar, N. A New Global Maximum Power Point Tracking Technique for Solar Photovoltaic (PV) System Under Partial Shading Conditions (PSC). *Energy* **2016b,** 512–525.

Ram, J. P.; Babu, T. S.; Rajasekar, N. A Comprehensive Review on Solar PV Maximum Power Point Tracking Techniques. *Renew. Sustain. Energy Rev.* **2017a,** *67*, 826–847.

Ren, H.; Wu, Q.; Gao, W.; Zhou, W. Optimal Operation of a Grid-connected Hybrid PV/Fuel Cell/Battery Energy System for Residential Applications. *Energy* **2016,** *113*, 702–712.

Sangeetha, K.; Babu, T. S.; Sudhakar, N.; Rajasekar, N. Modeling, Analysis and Design of Efficient Maximum Power Extraction Method for Solar PV System. *Sustain. Energy Tech. Assess.* **2016a,** *15*, 60–70.

Sangeetha, K.; Babu, T. S.; Rajasekar, N. Fireworks Algorithm-based Maximum Power Point Tracking for Uniform Irradiation as Well as Under Partial Shading Condition. In *Artificial Intelligence and Evolutionary Computations in Engineering Systems*; Springer: India, 2016b; pp 79–88.

Singh, S.; Singh, M.; Kaushik, S. C. Feasibility Study of an Islanded Microgrid in Rural Area Consisting of PV, Wind, Biomass and Battery Energy Storage System. *Energy Convers. Manag.* **2016,** *128*, 178–190.

CHAPTER 3

GRID INTEGRATION OF WIND TURBINES: ISSUES AND SOLUTIONS

A. RINI ANN JERIN*, K. PALANISAMY, and S. UMASHANKAR

School of Electrical Engineering, VIT University, Vellore, Tamil Nadu, India

Corresponding author. E-mail: riniannjerin@gmail.com

CONTENTS

ABSTRACT

The rising influence of the operation and control of wind energy on power systems has entailed grid integration as a key concern. Therefore, different countries are opting different approaches in setting up new grid codes in addressing this issue. Fault ride through (FRT) capability in wind turbines is established to maintain transient stability, which requires the wind generators to sustain the operation of the turbines during fault without tripping. The grid code specifications for FRT capability and the operation of FRT in different types of wind turbines are discussed in detail in this chapter. The hardware-based solutions available to provide FRT capability such as crowbar, DC-link chopper, static compensator, dynamic voltage restorer, magnetic energy recovery switch, energy storage systems, series grid side converter, fault current limiter are discussed in detail in this chapter.

3.1 INTRODUCTION

Wind is ubiquitous and wind energy is being harnessed extensively to fight the growing climate change and depleting fossil fuel threats (Li et al., 2015). Widespread adoption of wind energy in the pursuit of low-carbon economic growth, energy access, energy security, and prosperity has clearly marked that wind energy generation has grown extensively (Lopez et al., 2007). This growing grid integration of wind turbines has necessitated grid code modifications to accommodate the increasing wind energy and to provide seamless wind integration to the grid. The grid codes specify the need for transient stability of the wind turbines by provisioning fault ride through (FRT) capability which requires the wind generators to sustain the operation of the turbines during fault without tripping (Nasiri & Mohammadi, 2017). The grid integration of wind energy will allow the transfer of power generated in one area with high wind potential to be transmitted to other areas which have low or zero wind potential, thereby disseminating the wind energy generation for a wider region (Rashid & Ali, 2016). Although the modern developments in the wind technology have contributed to higher wind capture and power generation, the use of power electronic components have made the wind generators vulnerable to faults and other disturbances (Abdelrahem & Kennel, 2016).

Various techniques to augment the FRT capability in different types of wind turbines are proposed in the literature (Neves et al., 2016). The modern wind turbines installed after the grid code modifications have made the necessary extension in technology through hardware or software control to improve the FRT capability as per the grid code requirements (Tom et al., 2016). But there are several thousands of wind turbines which were installed previously before the establishment of FRT grid code requirements and require additional hardware devices based solutions. Therefore, this chapter extensively focuses on the hardware-based FRT solutions, widely available for enhancing the FRT capability of wind turbines (Ruiqi et al., 2016). The FRT capability and operation vary with each type of wind turbine and therefore, the different types of wind turbines which includes type 1—squirrel cage induction generator (SCIG), type 3—doubly fed induction generator (DFIG), and type 4—permanent magnet synchronous generator (PMSG) are discussed (Blaabjerg et al., 2006).

Hardware-based FRT solutions include the conventional crowbar, DC-link brake chopper method, shunt-based flexible alternating current transmission systems (FACTS) device such as static compensator (STATCOM), series-based FACTS devices such as the dynamic voltage restorer (DVR) and magnetic energy storage device (MERS), energy storage based methods, fault current limiter based method, and series grid side converter (SGSC) based FRT methods which are discussed in detail. The technical and economical study of these hardware-based solutions, their advantages and limitations, and feasibility are discussed in detail. This chapter will have a greater impact for the researchers focused in FRT solutions for previously installed wind turbines and in countries which are going through changes in the grid integration of wind turbines with the newly established grid code guidelines (Jain et al., 2015).

The organization of this chapter is as follows: Section 3.2 demonstrates the operation and FRT capability requirement in each type of wind turbine as per the grid codes; Section 3.3 discusses the different types of hardware-based FRT solutions and exemplifies the advantages and limitations of these solutions; Section 3.4 illustrates the FRT solutions based on the technical and economical advantages and feasibility; Section 3.5 gives a summary of the chapter with highlights. This chapter is intended to focus the benefits of the solutions proposed in literature with respect to the feasibility and performance features.

3.2   GRID CODES FOR FRT CAPABILITY

Grid codes are the specifications developed to regulate the wind farms according to the technical situations of the country. They may vary in scope and specification with respect to the need of the transmission system operators. The major grid codes are based on the active power, reactive power, voltage range, frequency operating range, and FRT capability requirements. Review on the grid codes is available in literature (Tsili & Papathanassiou, 2009), which has discussed the above specifications in detail.

During fault conditions, the voltage drops and this may lead to tripping of the generator during normal conditions. But FRT capability in wind turbine necessitates the wind turbine to remain connected to the grid for the time specified as per grid codes before disconnection. This ride through capability will ensure that there is no loss of power generation for normally cleared faults avoiding the negative impacts caused by disconnecting the wind turbine too quickly (Justo et al., 2015). This is particularly essential when a large capacity of wind power is grid connected, similar as in a wind farm (Tohidi & Behnam, 2016). Therefore, the FRT requirements in grid codes can be summarized as follows:

- Wind turbines are required to stay connected to the grid for a predefined period of time up to a certain level of voltage dip at the point of common coupling (PCC). The wind turbines are tested as per IEC 61400-21 standard to verify their operation under grid voltage disturbances. The curve shown in the grid code standard specifies that the wind turbine should stay connected to the grid above the curve level and disconnect if it goes below the curve as shown in Figure 3.1.
- Also, the wind turbines should contribute to the reactive support by generating reactive current during voltage dips. The reactive power curve as per German grid code as shown in Figure 3.2 shows that the wind turbine should produce 2% reactive current for each percent of voltage dip, for voltage dip between 10% and 50% from nominal voltage limit.
- The active power recovery should take place soon after fault clearance to support the grid frequency.

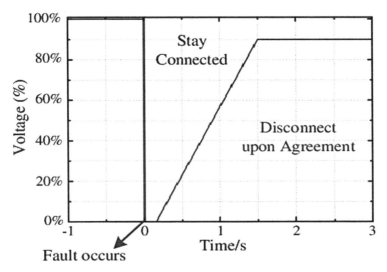

FIGURE 3.1   German grid code for fault ride through capability.

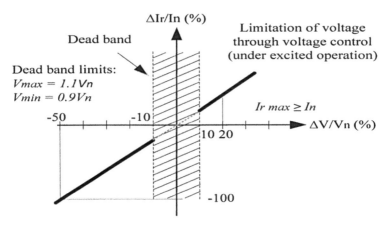

FIGURE 3.2   German grid code for reactive power support capability.

3.3   FRT CAPABILITY IN DIFFERENT TYPES OF WIND TURBINES

Wind turbines generally extract power from the wind by converting the wind energy into mechanical energy through the aerodynamic blades and later the mechanical energy into the electrical energy through the wind turbine rotor coupled to the generator through the gear box. The operation

and requirement of FRT solution are specific to the construction and operation of different types of wind turbine generator. A brief overview on the
basics of operation and control of each wind generator type is discussed.
In order to understand the operation of FRT capability during grid faults,
both the steady-state (normal operation) and transient-state (fault operation) of each generator type are discussed below and are also shown in
Figure 3.3.

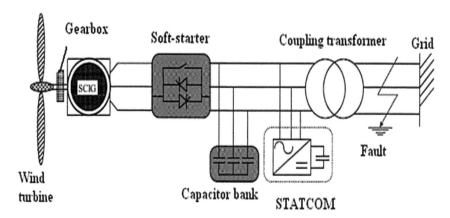

FIGURE 3.3   Squirrel cage induction generator-based wind turbine.

3.3.1   SCIG—TYPE 1

SCIG are the conventional fixed-speed wind turbines also termed as type 1
generator. They are the first generation wind turbines which were installed
widely then and are still in operation. They are still preferred in circumstances where low capacity wind turbines below 1 MW are incorporated
(Erlich et al., 2007). There are several grid-connected SCIG-based wind
turbines in operation which were installed before the advent of FRT capability requirement and grid code modifications and lack the FRT capability.
Also, the configuration of SCIG consists of a turbine rotor coupled to the
gearbox, the squirrel-caged generator, a soft-starter, mechanical-switched
capacitors and connected to the grid through a transformer. Since this type
of generator do not have any power electronic converters, the control of
the generator to maintain the transient stability cannot be done through
control and generally prefer external devices based FRT solutions. The

steady-state and normal operation of the SCIG-based generator is given below (Yaramasu et al., 2015).

3.3.1.1   STEADY-STATE OPERATION

During steady-state operation, when the wind speed exceeds the rated speed condition, the power delivered by the wind turbine may also exceed the rated value and this might cause damage to the wind turbines. Therefore, the aerodynamics is controlled by passive stall, active stall, and pitch control type techniques for SCIG-type low, medium, and large wind turbines, respectively. But the turbine is tripped down after the cutout speed value of the turbine to avoid system damage. Pitch control is used widely for the control of wind power extracted by controlling the wind blades. A capacitor bank is connected to the terminal of SCIG-based wind turbine in order to compensate the reactive power during steady-state operation.

The SCIG-based wind turbines equip induction generators due to its low cost and ruggedness which enables very low maintenance. But this includes certain drawbacks such as the stiff power requirement to enable steady-state operation of the generator, therefore requiring more pricey mechanical construction to absorb the excess mechanical stress. The wind gusts due to the erratic nature of the wind can cause torque pulsations on the drive train and may further lead to misalignment of the drive train and gradual gear component failure. This might gradually increase the operating cost and capital cost of the wind turbines leading to reduced paybacks.

Moreover, the wind turbines with SCIG-based generators have the general drawback of consuming huge reactive power, since the stator windings are connected to the grid directly without any power electronic interface. Therefore, bank of shunt capacitors is connected to the terminal of the wind generator to aid with the normal operation of the generator. This will also help the generator to achieve the unity power factor and voltage regulation requirements during steady-state operating conditions.

The induction generators also have the disadvantage of causing high inrush currents during starting of the operation and therefore utilize current limiters or soft-starters to avoid the disturbance caused to the grid and to eliminate the torque spikes to the drive train of the wind turbine. The soft-starters are generally thyristor-based technology which can limit the rms value of the inrush current to nearly twice the generator rated current value and also dampens the torque peaks to protect the drive train and reduces the load on the gear box.

3.3.1.2 TRANSIENT-STATE OPERATION

The transient operation of the wind turbine deals with the operation of the wind turbine during fault condition. The voltage drops at the PCC during fault, thereby leading to significant reduction in the electromagnetic torque and electric power output of the generator. But at the same time, the mechanical torque input of the generator is unaltered, thereby leading to abrupt rise in rotor speed beyond the limit to store the excess energy. Maintaining a balance between mechanical input energy and the electrical output energy is significant in improving the FRT operation of the SCIG-based generator types.

In SCIG-based wind generators, the rotor speed increases after fault clearance leading to reactive power absorption by the induction generator from the grid. This condition might exacerbate the voltage sag and increases the difficulty in restoring the terminal voltage within acceptable level. If output power (P_{out}) exceeds its rated value, the pitch angle (β) increases to limit the generated wind power to its rated value. This pitch angle control fails to operate effectively when the output power reduces during fault conditions. Thus, modified pitch angle controllers are proposed for the LVRT operation in SCIG to increase the pitch angle and thereby reduce the mechanical input torque. But the pitch angle control is very slow due to mechanical constraints; hence, we are required to look out for more advanced external controllers for enhancing the FRT capability of the wind turbines. Similarly, the capacitor banks utilized for reactive power compensation also provoke failures due to excessive switching and thereby increases the maintenance costs.

Although symmetrical faults are generally discussed, the majority of the events include the occurrence of asymmetrical faults. The unbalanced voltage in asymmetrical faults consists of both positive and negative sequence components, and therefore the stator current is unbalanced causing negative sequence currents. These negative sequence currents cause torque oscillations of double grid frequency causing the heating of stator windings, thereby reducing the life span of the gearbox, blade assembly, and other components.

$$\text{Positive sequence torque, } T^+ \cong 3 \times \frac{p}{2\omega_s} \times V_s^+ \times I_s^+ \qquad (3.1)$$

$$\text{Negative sequence torque, } T^- \cong 3 \times \frac{p}{2\omega_s} \times V_s^+ \times I_s^-, \qquad (3.2)$$

where p is the number of poles, ω_s is the sliding angular frequency, V_s^+ is the positive sequence voltage and I_s^+ and I_s^- are the positive and negative sequence currents, respectively.

The average torque reduces with decrease in positive sequence voltage and thereby increases the mechanical vibrations and noise. Negative sequence current injection to increase negative sequence voltage is proposed in literature.

3.3.2 DFIG—TYPE 3

DFIG-based wind turbines, also termed as type 3 wind turbines are very popular for their partial scale converter based configuration. They are widely used for wind energy conversion around the world. They are widely established and due to their power electronics based topology, they are very sensitive to grid side disturbances and require FRT support. Many control-based solutions have been proposed in the recent times, yet the wind turbines which are widely operating already require hardware-based solutions. The steady-state and transient operation of the DFIG and their configuration are discussed below, and the configuration is shown in Figure 3.4 (Abad et al., 2011).

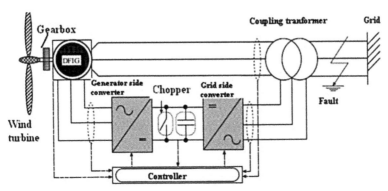

FIGURE 3.4 Doubly fed induction generator-based wind turbine.

3.3.2.1   STEADY-STATE OPERATION

The steady-state operation of DFIG employs d-q synchronous reference frame to regulate the DFIG-based wind turbines. During steady-state operation, the grid side converter controller controls the stator active power through q-axis rotor current component and the reactive power between stator and grid through the d-axis rotor current. Independent control of stator active and reactive power by using the rotor current regulation is achieved through this control.

The major objective of the grid side inverter is to keep the dc-link voltage constant regardless of the direction of power flow from the rotor. Reference frame oriented along stator or supply reference frame is utilized. The d-axis current is controlled to maintain the dc-link voltage constant and q-axis current is controlled to regulate the reactive power flow.

3.3.2.2   TRANSIENT-STATE OPERATION

DFIG-based generators are vulnerable to grid disturbances and therefore the voltage may drop at PCC causing large stator transient current. Due to magnetic coupling between stator and rotor windings, high voltages are induced in the rotor terminals and result in high current flow through the converters. Such transient current flow may easily damage the power electronic switches even in a very short interval of time. This happens mostly due to the small rating of the converters which is unable to produce the required voltage to control the generator during transient-state. The large stator voltage changes will cause large rotor voltage change due to the small converter rating. Therefore, large inrush currents are produced in the rotor windings and hence require a rotor current protection during fault conditions.

3.3.3   PMSG—TYPE 4

PMSG is also termed as a type 4 wind turbine and this utilizes a full variable speed generator with full capacity power electronic AC–DC–AC converter. Generally, PMSG-based wind turbines are preferred for high-power applications with power rating up to several megawatts. This type of wind generator is highly preferred for offshore wind applications due to the highest wind energy conversion efficiency offered by them. Due to the full-range converter, this generator can operate fully decoupled from the grid. They also offer smooth grid connection and can perform reactive power compensation. Since they employ full-scale converters, they can offer best FRT compliance through control methods without employing any external hardware-based solutions.

The major advantage of PMSG based wind turbines is the elimination of gearbox for high pole number and thereby reducing the cost of the machine, and the configuration is shown in Figure 3.5. Though this type of machine employs full-scale converter, it only amounts to less percent (approximately 7–12%) of the total cost of the machine. The PMSG-based generators are becoming popular in the recent days and hence are employed after the major grid code modifications for FRT. Therefore, most of the PMSG-based wind turbines employ FRT capability and are very effective in offering low voltage ride through during faults.

Hardware-based solutions are most essential for grid interconnection issues in PMSG-based generators for voltage stability at the PCC and active power recovery and reactive power support.

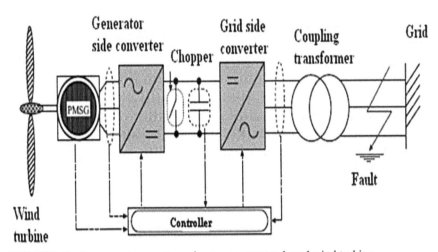

FIGURE 3.5 Permanent magnet synchronous generator-based wind turbine.

3.3.3.1 STEADY-STATE AND TRANSIENT OPERATION

The PMSG converter is similar to the DFIG converter with a larger capacity and includes generator side converter (GSC) control and grid side inverter

(GSI) control. The GSC controls the rotational speed of the PMSG for achieving the variable speed operation of the WECS, and the GSI controls the dc-link voltage and grid voltage. The d-q control based vector control strategy is used in the control methodology (Li, 2009). The PMSG is realized on rotating reference frame based speed control, where the rotational speed error is utilized as the reference of the speed control. In the inverter control system, the d-axis current controls the dc-link voltage and q-axis controls the grid voltage. The chopper circuit is included in the configuration of PMSG as shown in Figure 3.5., to avoid any damage to the dc-link capacitor due to large oscillations of the dc-link voltage during fault conditions. Also, high variations in active and reactive power during fault are avoided by the chopper circuit, to ensure the system stability (Tripathi et al., 2015).

3.4   FRT CAPABILITY USING HARDWARE-BASED SOLUTIONS

3.4.1   BLADE PITCH ANGLE CONTROL BASED LVRT

Pitch angle control adjusts the rotor speed by changing the pitch angle of the blade (β) to reduce the wind power extracted. Since the mechanical output power controls the angular speed, it is thereby used to control the output torque of the wind turbine. Pitch control is highly essential to protect the wind turbine from wind gusts. Control of power generated by the pitch control enables frequency control operation and thereby contributes to the power stabilization. During fault condition, the pitch control is adjusted to increase the blade pitch angle and to reduce the mechanical power extraction.

Modified GSC control of DFIG can convert the additional power into WT kinetic energy instead of dissipating through the crowbar resistance which temporarily increases the generator rotor speed during grid faults. This reduces the oscillations in currents and also the pitch control is triggered when the rotor speed exceeds the rated speed. Limiting the rotor speed will prevent the mechanical stress overload to the turbine system. But pitch control being mechanical system has the limitations in speed and therefore require combined efforts of the advanced converter control to enhance the FRT capability through this method (Tripathi et al., 2015). The modified pitch angle controller used in wind turbines is shown in Figure 3.6.

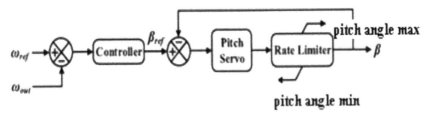

FIGURE 3.6   Modified pitch angle controller.

3.4.2   CROWBAR METHOD

Crowbar is the most well-established protection circuit based technique which acts similar to a dump load. It creates a low resistance path during fault conditions, thereby isolating the converters and preventing it from overcurrents. It is connected across the power electronic switching devices of the GSC and the slip rings.

Conventional topology of crowbars includes antiparallel thyristor-based crowbar and diode bridge crowbar. The advancements in crowbar have led to active crowbar topology which eliminates the short circuit current at any instant. Diode bridge crowbar is an optimal topology for active crowbar-type protection. For successful ride through of faults, the short circuit made by the crowbar has to be removed before the operation of the converters (Li & Zhang, 2010). Premature removal of the crowbar does not serve the purpose of converter protection and late removal leads to higher absorption of reactive power from the grid.

The appropriate value of the crowbar resistance can be selected by using mathematical analysis (Li-Ling et al., 2012). It is generally chosen as a compromise between certain objectives based on the performance of DFIG and the compliance of grid codes. Very low resistance value leads to very large currents during the voltage dip and a large value causes big peak in electromagnetic torque. Higher resistance reduces the rotor current but cannot reduce rotor voltage. Therefore, to increase the dc bus voltage, the rotor current circulates even when inactive via the freewheeling diodes. Therefore, based on the mathematical analysis and simulations, an appropriate crowbar resistance with the safety of the machine prioritized is chosen. This is considered for minimizing the torque and rotor current peaks during dips.

The temporary loss of GSC control leads to reactive power absorption which escalates the voltage dip and postpones grid recovery. Crowbar protection is also recognized as rotor current limiter (RCL) (Hossain et al., 2013). The series dynamic breaker (SDR) is connected to the rotor side of generator in order to limit the rotor overcurrent. A combination of SDR and conventional crowbar gives the advantage of uninterrupted operation of DFIG control during the crowbar protection activation. The crowbar protection used for FRT capability in a DFIG-based wind turbine is shown in Figure 3.7.

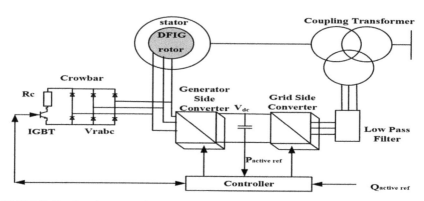

FIGURE 3.7   Crowbar protection for FRT capability improvement in DFIG-based generator.

3.4.3 DC-LINK BRAKE CHOPPER METHOD

A chopper circuit is a resistance added to the dc-link side to reduce the dc-link voltage and acts similar to rotor side crowbar. The chopper increases voltage at converter terminals enabling speedy dc-link voltage recovery. The dc-link brake chopper alone employed for FRT capability of DFIG wind turbines with two different control methods is shown by Gong et al. (2010) and the results are compared with the conventional crowbar method and found to be more effective.

In chopper method, the time taken for converter disengagement and restoration is longer than the crowbar control, since it does not assist demagnetization of the electrical machine post fault. Therefore, the electrical performance of a dc-link brake chopper is quite inferior when compared to crowbar. But crowbar can reduce the rotor decay time with suitable rotor resistance selection which assists faster control recovery

(Kawady et al., 2010). Yet, a dc-link chopper is a valuable choice for lower values of dc-link capacitance, which are sensitive to rotor overcurrents.

3.4.4 STATIC COMPENSATOR

STATCOM was introduced in a type 1 generator system to stabilize and recover faults in a large scale wind farm (Kamel, 2014). STATCOMs are connected between the PCC of the wind farm and the grid to perform the FRT operation. During steady-state, STATCOM will inject/absorb reactive power to safeguard the bus voltage and prevent fluctuations. During transients, in order to hasten the voltage recovery and reestablish the stability, the STATCOM will inject maximum reactive power. The configuration of STATCOM connected to wind turbine is shown in Figure 3.8.

But for type 3 DFIG-based generators, the operation and control of STATCOM has a different structure. The grid side converter of the DFIG itself can operate similar to a STATCOM and supply the reactive power to support the fault ride through, while the GSC is disabled by crowbar. STATCOMs are not applicable to prevent the faults occurring within the wind farm and hence cannot provide complete protection. Installation of big STATCOMs to ride through fault in wind farms has been discouraged due to high costs (Gonzelez & Guixot, 2010). Real time implementation of STATCOM in a DFIG-based wind farm shows that STATCOM provides dynamic voltage support and re-establishes the voltage at PCC shortly after grid fault.

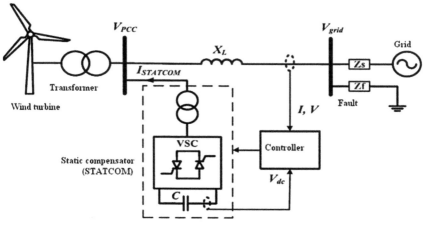

FIGURE 3.8   FRT capability of wind turbine utilizing STATCOM.

3.4.5 DYNAMIC VOLTAGE RESTORER

DVR is a series compensating device to ride through the wind turbine during grid faults through voltage control (Ramirez et al., 2011). It consists of a voltage source converter for voltage injection, connected between the wind turbine and grid through a coupling transformer, which has an energy source and includes filters for harmonic elimination. DVR configuration is similar to that of the static synchronous series compensator (SSSC) with direct control over the terminal voltage using capacitor bank or energy storage device. It has an additional voltage source converter (VSC) employed at generator terminals to perform the series voltage compensation.

Even though the application of DVR for FRT capability of DFIG is expensive, it is capable of eliminating the transients in generator currents and power at grid fault conditions effectively. FRT capability of SCIG with DVR is shown by Ramirez et al. (2011). FRT capability of DFIG with DVR for balanced faults is shown by Ibrahim et al. (2009), without considering reactive power and transient fault conditions. FRT capability of DFIG with DVR for unbalanced faults is shown by Wessels et al. (2011) and transient grid conditions are included.

Decreasing the stator power reference for abnormal grid voltages reduces the DVR power ratings significantly. The schematic of the DVR connected to DFIG and grid is shown in Figure 3.9. Generally, the series compensation employing DVR is carried out in four different schemes based on the requirement: in-phase, pre-sag, minimum energy and zero-active power injection based compensation.

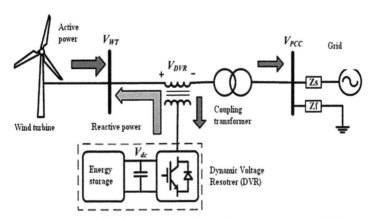

FIGURE 3.9   FRT capability by employing dynamic voltage restorer (DVR).

Series compensators with reduced power capacity are observed to be much more effective in restoring voltage compared to the parallel reactive power compensator in strong grid utility (Zhang et al., 2012). Since DVR avoids the need for any other protection equipment necessary for DFIG, it is a better solution for wind turbines installed already and to protect any distributed load in a microgrid (El Moursi et al., 2014).

3.4.6 MAGNETIC ENERGY RECOVERY SWITCH (MERS)

Magnetic Energy Recovery Switch (MERS) consists of four power electronic switches and a capacitor similar to single phase full bridge converter. The arrangement has two of the converter terminals connected in series provides series compensation during voltage sag due to faults. MERS was first developed in Tokyo Institute of Technology's Shimada Laboratory (Wiik et al., 2007). MERS as a LVRT capability solution for SCIG-based wind turbines was proposed by Mahvash and Taher (2016). The configuration of MERS connected to wind turbine for improving FRT capability is shown in Figure 3.10.

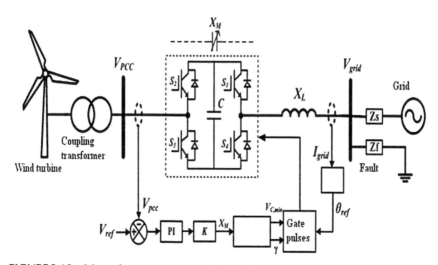

FIGURE 3.10   Magnetic energy recovery switch (MERS)-based FRT capability improvement.

The device creates some harmonics in line current whose effects are not severe, but causes interference with the resonance frequency of the system to which it is applied and further study is required to avoid this disturbance. Even though MERS is classified as a viable FACTS controller for LVRT capability of induction generators (Cheng et al., 2015), it is still not studied widely for DFIG application. MERS is also studied in combination with PMSG-based WTs and are found to offer a comparatively simple rectifier with reactive power controllability and advantageous in high synchronous reactance case.

3.4.7 ENERGY STORAGE BASED METHOD

Energy storage based FRT capability of wind turbines is achieved through reactive power support to the grid, to protect the dc-link of the converters from over-voltages and can regulate the active power output of the generator. Energy storage can improve the transient dynamics of DFIG by controlling rotor current (Abbey & Joos, 2007) and the power systems transient stability (Shen et al., 2016). Also, it includes other benefits such as frequency regulation, enhancement of voltage stability, participation of unit commitment, and electricity market operation (Díaz-González et al., 2012). The transient EMF is reduced by injecting the demagnetizing current from the converters to the rotor circuit for enhancing LVRT capability of DFIG. Among the several types of ESS devices, batteries and supercapacitors react faster during transients.

The capacity of the ESS and the current rating of the supplementary converter are determined optimally. Fast power modulation is provided by high ramp-up rates possible for batteries, flow batteries, supercapacitors (Biao et al., 2011), flywheels, and superconducting magnetic energy storage (SMES) (Riouch et al., 2015). The dc-link capacitors complemented by supercapacitors provide effective FRT support.

Vanadium redox-flow battery (VRB) based batteries are also used to improve the generator capability during ride through by absorbing the excessive energy of dc-link capacitor of PMSG for the LVRT capability enhancement. A battery system can reduce the power fluctuations due to wind speed changes and limits the excess power via GSC through the dc-link during fault. BESS control strategy is used to mitigate symmetrical voltage dips to aid the LVRT capability of DFIG by providing reactive power support during low voltage conditions by utilizing a flywheel to act similar to a STATCOM (Fathima & Palanisamy, 2015). Therefore, this approach has recorded to effectively mitigate oscillations, reduces stress on protective devices, and increases wind energy penetration.

3.4.8 SERIES GRID SIDE CONVERTER

SGSC is a second VSC connected across the dc-link. The output voltage of SGSC is regulated to control the stator terminal voltage to aid the DFIG-based generator to overcome deep voltage sags. It can also reduce or eliminate the negative sequence flux and transient dc. The stator voltage unbalances due to negative sequence voltage causes stator and rotor unbalance and electromagnetic torque and power pulsations in DFIG. Therefore, when it is eliminated, the positive sequence voltage will be left. This will naturally eliminate the unbalance in DFIG (Heller & Schumacher, 1997).

The SGSC was first promoted by Ran et al. (2006), but it does not include any in-depth studies of its properties and limitations. Further exploration of SGSC has promised excellent potential for FRT capability but shortcomings in power processing capability (Morren & de Haan, 2005). SGSC is also capable of coping with long term steady-state grid voltage unbalances. SGSC has good operational characteristics, but the economic viability needs to be analyzed for practical implementation (Flannery & Venkataramanan, 2008).

3.4.9 FAULT CURRENT LIMITER

The high penetration of wind energy has led to high fault current levels during grid faults. Therefore, fault current limitation using fault current limiters (FCL) is employed. There are different types of FCL, but bridge-type FCL and superconducting fault current limiters (SFCL) are widely utilized. SFCL can limit the fault currents by automatically adding a nonlinear resistance which transits from superconducting state to the normal state. The main advantage is that they do not add any impedance to the system during normal operations.

FCL can be utilized to limit the rotor side overcurrents in DFIG converters for protection during FRT operation. The advancements in FCL have led to the utilization of solid-state fault current limiter (SSFCL) as shown by Zou et al. (2016). These SSFCL can be categorized as bridge, resonant, and switch type. The cost of switch-type fault current limiter

(STFCL) is higher when compared to crowbar, but has negligible on-state losses of semiconductor devices. STFCL can limit fault current, rotor back-EMF voltage and also has enhanced RSC controllability. Therefore, provides outstanding LVRT enhancing capability. When stator side is chosen for installation of SFCL, DFIG's terminal voltage can be improved to prevent the disconnection of DFIG during fault, and this enhances the operational stability. Resistive-type SFCL utilized for FRT capability operation in fixed speed wind turbine is shown in Figure 3.11, where the current limiting is modeled by the resistance transition in terms of temperature and current density.

In the work done by Ngamroo (2016), SFCL in combination with high power density SMES is also utilized, since SMES can be opted as an ideal solution for high power applications. The SFCL-MES solution proposed in (Guo et al., 2012), utilizes the GSC of DFIG along with an isolation transformer, diode rectifier and chopper as Energy Conservation System (ECS) to perform the output power smoothing and LVRT capability enhancement together.

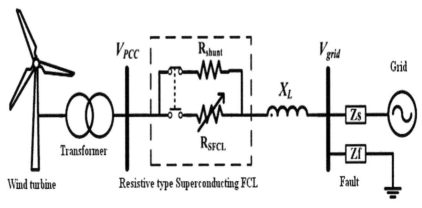

FIGURE 3.11   FRT capability improvement using resistive type SFCL.

3.5   CONCLUSION

FRT capability in wind turbines has received utmost importance due to the steady growth in grid integration of wind energy. This chapter discusses the grid code requirement for FRT capability, the steady-state and transient operation of the different types of wind turbines which includes fixed s-speed (SCIG), variable-speed (DFIG and PMSG) based generators. The FRT capability requirements in the different types of generators are discussed and based on which the various hardware-based solutions available for the FRT capability augmentation is elaborated. The conventional crowbar-based solution, blade pitch angle control, dc-link brake chopper, STATCOM, DVR, MERS, SGSC, energy storage based solutions, FCL are discussed briefly. This chapter gives an overview on the different hardware-based solutions that can be commonly applied for the different types of wind generators and the advantages and disadvantages offered by the solutions. Some of the solutions discussed are at their nascent stage and are under development for commercial viability. Technical and economical constraints may alter the choice of the solution. FACTS devices-based solutions are very popular for their performance and fault current limiting is required for protection of the wind energy conversion devices. Thus, FRT capability can be effectively met through the above discussed solutions.

KEYWORDS

- renewable energy
- fault ride through
- low voltage ride through
- wind turbines
- grid codes

REFERENCES

Abad, G., et al. *Doubly Fed Induction Machine: Modeling and Control for Wind Energy Generation,* 1st ed.; Wiley Publications, 2011; pp 144 (Chapter 2).

Abbey, C.; Joos, G. Supercapacitor Energy Storage for Wind Energy Applications. *IEEE Trans. Ind. Appl.* **2007,** *43*(3), 769–776.

Abdelrahem, M.; Kennel, R. Fault-ride Through Strategy for Permanent-magnet Synchronous Generators in Variable-speed Wind Turbines. *Energies* **2016,** *9*(12), 1066.

Biao, M., et al. In *Studies on Security Capacity of Wind Farms Containing VRB Energy Storage System*, DRPT 2011, 4th International Conference on Electric Utility Deregulation and Restructuring and Power Technologies, 2011; pp 1704–1708.

Blaabjerg, F., et al. Overview of Control and Grid Synchronization for Distributed Power Generation Systems. *IEEE Trans. Ind. Electron.* **2006**, *53*,1398–409.

Cheng, M. M., et al. Characteristics of the Magnetic Energy Recovery Switch as a Static Var Compensator Technology. *IET Power Electron.* **2015**, *27*(13), 29–38.

Díaz-González, F., et al. A Review of Energy Storage Technologies for Wind Power Applications. *Renew. Sustain. Energy Rev.* **2012**, *16*(21), 54–71.

El Moursi, M. S.; Goweily, K.; Kirtley, J. L.; Abdel-Rahman, M. Application of Series Voltage Boosting Schemes for Enhanced Fault Ride Through Performance of Fixed Speed Wind Turbines. *IEEE Trans. Power Deliv.* **2014**, *29*, 61–71.

Erlich, I.; Wrede, H.; Feltes, C. In *Dynamic Behavior of DFIG-based Wind Turbines During Grid Faults*, Fourth Power Convers. Conf. PCC-NAGOYA 2007, Conf. Proc., 2007; pp 1195–1200.

Fathima, A. H.; Palanisamy, K. Optimization in Microgrids with Hybrid Energy Systems—A Review. *Renew. Sustain. Energy Rev.* **2015**, *45*, 431–446.

Flannery, P. S.; Venkataramanan, G. A Fault Tolerant Doubly Fed Induction Generator Wind Turbine Using a Parallel Grid Side Rectifier and Series Grid Side Converter. *IEEE Trans. Power Electron.* **2008**, *23*(3), 1126–1135.

Gong, B.; Xu, D.; Wu, B. In *Cost Effective Method for DFIG Fault Ride-through During Symmetrical Voltage Dip*, IECON Proc. (Industrial Electron Conf.), 2010; pp 3269–3274.

Gonzelez, J. I.; Guixot, M. V. (Inventors; Gamesa Innovation; Technology, S. L.; Assignee). Method and Device for Injecting Reactive Current During a Mains Supply Voltage Dip. U.S. Patent 7,821,157, 2010.

Guo, W.; Xiao, L.; Dai, S. Enhancing Low-voltage Ride-through Capability and Smoothing Output Power of DFIG with a Superconducting Fault-current Limiter–Magnetic Energy Storage System. *IEEE Trans. Energy Convers.* **2012**, *27*(2), 277–295.

Heller, M.; Schumacher, W. In *Stability Analysis of Doubly-fed Induction Machines in Stator Flux Reference Frame*, European Conference on Power Electronics and Applications, 1997; Vol. 2, pp 2–707.

Hossain, M. J., et al. Control Strategies for Augmenting LVRT Capability of DFIGs in Interconnected Power Systems. *IEEE Trans. Ind. Electron.* **2013**, *60*, 2510–2522.

Ibrahim, A. O.; Nguyen, T. H.; Lee, D. C.; Kim, S. C. In *Ride-through Strategy for DFIG Wind Turbine Systems Using Dynamic Voltage Restorers*. Energy Conversion Congress and Exposition, 2009. ECCE 2009,California, USA, 2009, 1611–1618.

Jain, B.; Jain, S.; Nema, R. K. Control Strategies of Grid Interfaced Wind Energy Conversion System: An Overview. *Renew. Sustain. Energy Rev.* **2015**, *47*, 983–996.

Justo, J. J.; Mwasilu, F.; Jung, J. W. Doubly-fed Induction Generator Based Wind Turbines: A Comprehensive Review of Fault Ride-through Strategies. *Renew. Sustain. Energy Rev.* **2015**, *45*, 447–467.

Kamel, R. M. Three Fault Ride Through Controllers for Wind Systems Running in Isolated Micro-grid and Effects of Fault Type on Their Performance: A Review and Comparative Study. *Renew. Sustain. Energy Rev.* **2014**, *37*, 698–714.

Kawady, T., et al. In *Protection System Behavior of DFIG Based Wind Farms for Grid-faults with Practical Considerations*, IEEE PES Gen. Meet. PES, 2010.

Li, S. Characteristic Study of Vector-controlled Direct-driven Permanent Magnet Synchronous Generator in Wind Power Generation. *Electr. Power Compon. Syst.* **2009**, *37*(10), 1162–1179.

Li, D.; Zhang, H. In *A Combined Protection and Control Strategy to Enhance the LVRT Capability of a Wind Turbine Driven by DFIG*, 2nd Int. Symp. Power Electron Distrib. Gener. Syst. PEDG, 2010; pp 703–707.

Li, P., et al. Control and Monitoring for Grid-friendly Wind Turbines: Research Overview and Suggested Approach. *IEEE Trans. Power Electron.* **2015**, *30*, 1979–1986.

Li-Ling, S.; Pu, Y.; Yi, W. In *Simulation Research for LVRT of DFIG Based on Rotor Active Crowbar Protection*, Sustain. Power Gener. Supply (SUPERGEN 2012) Int. Conf., 2012; pp 1–7.

Lopez, J., et al. Dynamic Behavior of the Doubly Fed Induction Generator During Three-Phase Voltage Dips. *IEEE Trans. Energy Convers.* **2007**, *22*, 709–717.

Mahvash, H.; Taher, S. A. A Look-up Table Based Approach for Fault Ride-through Capability Enhancement of a Grid Connected DFIG Wind Turbine. *Sustain. Energy Grids Networks* **2016**.

Morren, J.; de Haan, S. W. H. Ridethrough of Wind Turbines with Doubly-fed Induction Generator During a Voltage Dip. *IEEE Trans. Energy Convers.* **2005**, *20*, 435–441.

Nasiri, M.; Mohammadi, R. Peak Current Limitation for Grid Side Inverter by Limited Active Power in PMSG-Based Wind Turbines During Different Grid Faults. *IEEE Trans. Sustain. Energy* **2017**, *8*(1), 3–12.

Neves, F. A., et al. Unbalanced Grid Fault Ride-through Control for Single-stage Photovoltaic Inverters. *IEEE Trans. Power Electron.* **2016**, *31*(4), 3338–3347.

Ngamroo, I. Optimization of SMES-FCL for Augmenting FRT Performance and Smoothing Output Power of Grid-connected DFIG Wind Turbine. *IEEE Trans. Appl. Supercond.* **2016**, *26*(7), 1–5.

Ramirez, D., et al. Low-voltage Ride-through Capability for Wind Generators Based on Dynamic Voltage Restorers. *IEEE Trans. Energy Convers.* **2011**, *26*, 195–203.

Ran, L. R. L., et al. In *Control of a Doubly Fed Induction Generator in a Wind Turbine During Grid Fault Ride-through*, IEEE Power Eng. Soc. Gen. Meet., 2006; Vol. 21, pp 652–662.

Rashid, G.; Ali, M. H. In *Asymmetrical Fault Ride Through Capacity Augmentation of DFIG Based Wind Farms by Parallel Resonance Fault Current Limiter*, IEEE Power and Energy Society General Meeting (PESGM), 2016; pp 1–5.

Riouch, T.; Alamery, A.; Nichita, C. Control of Battery Energy Storage System for Wind Turbine Based on DFIG During Symmetrical Grid Fault, La Coruna, Spain, 2015, 1–5.

Ruiqi, L. I.; Hua, G.; Geng, Y. Fault Ride-through of Renewable Energy Conversion Systems During Voltage Recovery. *J. Mod. Power Syst. Clean Energy* **2016**, *4*(1), 28–39.

Shen, Y., et al. Advanced Auxiliary Control of an Energy Storage Device for Transient Voltage Support of a Doubly Fed Induction Generator. *IEEE Trans. Sustain. Energy* **2016**, *7*, 63–76.

Tohidi, S.; Behnam, M. A Comprehensive Review of Low Voltage Ride Through of Doubly Fed Induction Wind Generators. *Renew. Sustain. Energy Rev.* **2016**, *57*, 412–419.

Tom, P. M., et al. A Technical Review on LVRT of DFIG Systems. In *Information Systems Design and Intelligent Applications;* Springer: New Delhi, 2016; pp 397–404.

Tripathi, S. M., et al. Grid-integrated Permanent Magnet Synchronous Generator Based Wind Energy Conversion Systems: A Technology Review. *Renew. Sustain. Energy Rev.* **2015**, *51*, 1288–1305.

Tsili, M.; Papathanassiou, S. A Review of Grid Code Technical Requirements for Wind Farms. *IET Renew. Power Gener.* **2009,** *3*, 308.

Wessels, C.; Gebhardt, F.; Fuchs, F. W. Fault Ride-through of a DFIG Wind Turbine Using a Dynamic Voltage Restorer During Symmetrical and Asymmetrical Grid Faults. *IEEE Trans. Power Electron.* **2011,** *26*, 807–815.

Wiik, J. A.; Wijaya, F. D.; Shimada, R. An Innovative Series Connected Power Flow Controller, Magnetic Energy Recovery Switch. **2007,** 1–7.

Yaramasu, V., et al. High-power Wind Energy Conversion Systems: State-of-the-art and Emerging Technologies. *Proc. IEEE* **2015,** *103*, 740–788.

Zhang, S., et al. Advanced Control of Series Voltage Compensation to Enhance Wind Turbine Ride Through. *IEEE Trans. Power Electron.* **2012,** *27*, 763–772.

Zou, Z. C., et al. Integrated Protection of DFIG-Based Wind Turbine with a Resistive-Type SFCL Under Symmetrical and Asymmetrical Faults. *IEEE Trans. Appl. Supercond.* **2016,** *26*(7), 1–5.

CHAPTER 4

CONTROL STRATEGIES FOR RENEWABLE ENERGY SYSTEMS

RAMJI TIWARI[1], K. KUMAR[2], and N. RAMESH BABU[3*]

[1,2]*School of Electrical Engineering, VIT University, Vellore, Tamil Nadu 632014, India*

[3]*M. Kumarasamy College of Engineering, Karur, Tamil Nadu, India*

Corresponding author. E-mail: nrameshme@gmail.com

CONTENTS

ABSTRACT

Renewable energy system is growing exponentially for continuous increase in energy demand and depletion of fossil fuels. The renewable energy has a great potential in terms of producing clean energy and reducing the carbon emission which is generated from the other sources. Renewable energy such as solar and wind are the emerging alternative for fossil fuel-based generation and is expected to play a significant role in producing clean energy throughout the world to achieve zero carbon footprint. The growing demand of renewable energy tends to produce a quality output power and capable to synchronize the exiting grid within the standards. An appropriate controller is required to control the power produce by the different renewable energy sources since they are highly intermittent in nature. In this chapter, a concise review of a different control strategies associated with the solar- and wind-based generation is presented. Thus, this chapter is intended to provide a suitable reference for further research in the field of determining the control strategies for renewable-based generation using solar, wind as well as hybrid power generation.

4.1 INTRODUCTION

Renewable energy-based power generation is rapidly growing across the world to overcome the shortage of the energy demand. At present, the major source of energy is through fossil fuel. The constant depletion of fossil fuel, which is about to extinct, has forced the energy sector to concentrate on the alternative source of energy (Tiwari et al., 2016). The solar- and wind-based renewable energy is mainly focused since they are widely available. The reliability on the solar and wind highly depends on the climatic changes and unpredictable nature (Moretti et al., 2016). Thus, an efficient control strategy should be implemented along with the renewable energy to provide better stability for grid integration (Eltigani & Masri et al., 2015).

This chapter deals with the various control strategies for solar and wind energy. It also includes the suitable control operations when the solar and wind energies are integrated to form a hybrid system. The conventional and innovative control strategies which are employed in literature are explained. The unpredictable nature of renewable energy tends to improvise the control strategy in order to achieve more reliable and sustainable energy.

The control strategies of solar-based power generation are explained in Section 4.2, where the maximum power point tracking (MPPT), current-based control technique, grid side controller (GSC), and fault ride through (FRT) control techniques are detailed and a brief summary is reported.

Section 4.3 of this chapter constitutes of control strategies of wind energy. The control techniques which are associated with wind-based power generation are pitch angle controller, MPPT controller, grid side controller, and low voltage ride through (LVRT) controller are explained. The control technique based on the offshore grid connected wind energy conversion is also precisely summarized.

The control techniques used to mitigate the power quality issues in hybrid renewable energy (HRE) are explicitly described in Section 4.4. The control strategies in HRE are mainly implemented to provide a steady supply to the grid. The future development of HRE is explained in this section.

4.2 CONTROL STRATEGIES IN THE SOLAR SYSTEM

The impact of renewable energy such as solar energy has increased during oil crisis. In earlier years when the oil price was cheaper and the cost of renewable energy was affecting the economic balance, the interest in solar-based power generation was minimal (Robert et al., 2003). Since the extinct of fossil fuel is felt and the impact on environmental issue is studied, the renewable-energy-based generation has been in focus in major countries (Abbasi et al., 2011). The investment in solar energy has increased manifold and the cost of manufacturing the solar cell has also decreased by 70% in 2010 compared to previous year (Branker et al., 2011). The major issues faced by the solar energy are: the efficient way to utilize the solar radiation, convert the solar energy into electrical energy, store the energy obtained, and to reduce the overall cost of the system (Singh, 2013). The major drawbacks of solar energy are to provide the energy to the consumer at affordable cost and intermittent availability of solar energy (Bazilian et al., 2013).

The above drawbacks can be overcome by using effective way to control the parameters of solar energy. The considerable research is focused on the way to improve the control strategies of the solar system to provide efficient and uninterrupted power to the consumer end. Advance control strategies which are capable to cope with the dynamic, nonlinearity of the solar radiation are briefly explained.

The complete block diagram of solar-based power generation technique is illustrated in Figure 4.1. The converter and inverter play a vital role in controlling the entire system. The DC/DC converter is the main component in electrical system of solar conversion. The maximum power point control is implemented in this converter. The grid side controller controls the inverter system. The control methods used in solar-based energy conversion system are MPPT, grid synchronization, and power quality control, which are explained in detail.

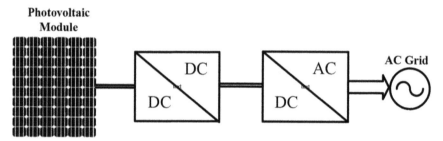

FIGURE 4.1   Basic solar-based conversion system.

4.2.1   MAXIMUM POWER POINT TRACKING

MPPT is the most essential part of the solar system for improving the total energy harvesting. A vast research in tracking the optimum power obtained from available sun intensity has been carried out in recent years. Several MPPT control strategies are being implemented by the researchers and manufactures based on the parameters such as complexity, convergence time, cost, and stability of the system. Various MPPT techniques such as Perturb & Observe (P&O), incremental conductance (INC), and soft computing techniques such as fuzzy logic and neural networks have been implemented in literature to enhance the efficiency of solar system (Saravanan & Babu, 2016).

The power–voltage and current–voltage characteristics of photovoltaic (PV) panel are shown in Figure 4.2. The MPPT algorithm is intended to track the maximum voltage V_{MPP} or the maximum current I_{MPP} to obtain maximum power P_{MPP} for the available temperature and irradiance (Karami et al., 2017). The selection of MPPT control strategies are mainly based on the cost of implementation, electronic equipment and sensor requirement, speed of tracking the peak point, and complexity in operation which is specified in Table 4.1.

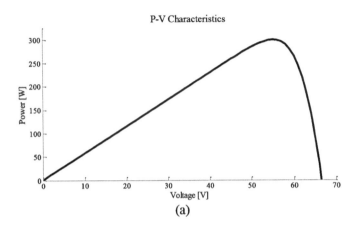

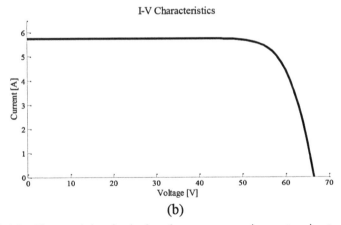

FIGURE 4.2   Characteristic of solar-based energy conversion system in standard test condition.

The basic MPPT also, such as P&O and INC, are based on the principle of hill-climb search approach (Messalti et al., 2017). They are most basic MPPT control technique used in the literature. They are widely used for its simplicity and cost-effective nature. The main idea for hill-climb search method is to adjust the duty cycle of the converter to provide suitable

TABLE 4.1 Comparison of Different MPPT Strategy for PV System.

MPPT technique		Cost	Parameter sensed	Speed	Complexity	Stability	Periodic tuning	Efficiency
Conventional methods	P&O	Low	Voltage, current	Slow	Less	Not stable	No	Low
	INC	Low	Voltage, current	Slow	Medium	Stable	No	Medium
Soft computing controller	FLC	Low	Voltage, current	Medium–fast	Less	Very stable	No	High
	ANN	Medium	Depends	Fast	High	Very stable	No	High
Hybrid controller		High	Depends	Depends	Varies	Very stable	Depends	Very high

power output based on available input (Subudhi & Pradhan, 2013). The P&O method determines the voltage difference between the PV panel and DC output of converter and tunes the duty cycle of the switch in the converter to yield maximum power (Jain & Agarwal, 2007). The latest obtained power of the perturb duty cycle is observed and compared with the previous step power. The difference in the power decides the increment and decrement of the duty cycle based on the need of the system as shown in flowchart in Figure 4.3. The variations in voltage and current are measured instantly in regular interval of time. If the variation of measured power is in a positive manner, the duty cycle is increased. If the obtained power is less than that of previous, the duty cycle is decremented. This algorithm repeats until the system reaches MPP. The major disadvantage of the system is the convergence speed. This method is mostly used for small scale. The oscillation in the MPP region causes issue in the stability of the system.

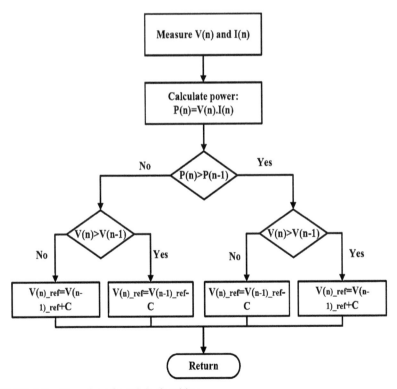

FIGURE 4.3 Flowchart for P&O algorithm.

Incremental conductance method is similar to that of P&O. Both the technique follows the same principle. The only difference is that in INC method the duty cycle is kept constant once it reaches the MPP until there is any considerable change in the current of the PV. It is based on the reference voltage (Safari & Mekhilef, 2011). The MPP is fixed by taking the difference between the generated voltage and reference voltage as illustrated in flowchart in Figure 4.4. The P&O and INC method cannot track the MPP during rapid variation in the solar intensity and they provide low convergence speed. To overcome the above method, soft computing method-based MPPT technique is implemented.

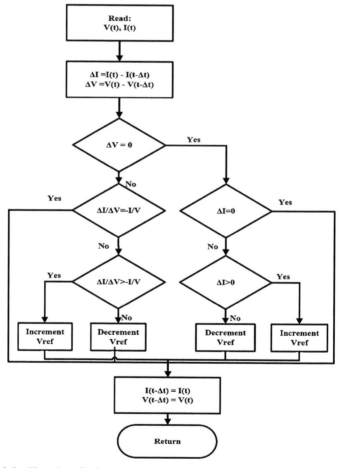

FIGURE 4.4   Flowchart for incremental conductance algorithm.

The fuzzy logic- and neural network-based MPPT techniques are implemented to generate precise output during the rapid variation in the atmospheric changes. The convergence speed of the soft computing-based MPPT method is far better than the basic MPPT controller (Saravanan & Babu, 2016). The oscillations at MPP are reduced to provide better stability to the solar PV system.

The fuzzy logic control (FLC) uses membership function for its operation instead of mathematical model. FLC control strategies consist of three stages: fuzzification, rule base inference system, and defuzzification as shown in Figure 4.5. The input variables are converted into linguistic variable in fuzzification process. The rules are framed according to the need of system in inference system. The linguistic variable is converted back to the numerical variable as an output variable in defuzzification process (Kwan & Wu, 2016). Normally, solar system has two inputs and one output. The choice of the input and output variable are based on users own choice. Basically, the error in power obtained and change in error are used as input variable and duty cycle is chosen as output variable for MPPT control. The duty cycle obtained for the FLC control strategy is given as the input to the converter to obtain the maximum power. FLC-based MPPT strategy provides fast convergence speed and tracks the MPP during rapid variation of the solar intensity. However, FLC strategy requires preknowledge of the system. Thus, the efficiency of the system is determined based on the approximation understanding about the system (Patcharaprakiti et al., 2005).

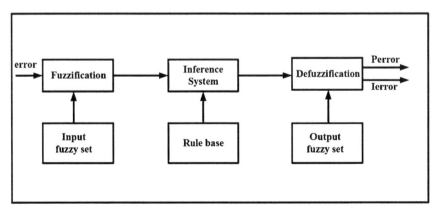

FIGURE 4.5   Fuzzy logic-based MPPT controller.

The neural network-based MPPT technique provides faster tracking than any of MPPT technique for determining precise MPP. This NN-based control strategy uses parameter approximation technique to overcome the nonlinearity of the system (Reisi et al., 2013). The NN technique consists of three layers: input layer, hidden layer, and output layer as shown in Figure 4.6. The input parameter for the NN-based MPPT technique is voltage and current from PV panel and the output is the duty cycle for the converter. The hidden layer is used to propagate the input signal to the output signal based on the transfer function applied to it. The determination of number of neurons in the hidden layer is chosen based on the trial and error method (Ishaque & Salam, 2013). The basic transfer function used for the MPPT technique is the tangent sigmoid for hidden layer and pure linear for output layer. The major advantage of the NN method is that it can produce more precise duty cycle under rapid variation of atmospheric changes and without having much knowledge of PV system.

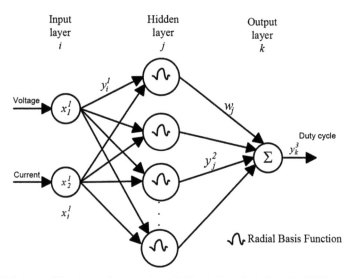

FIGURE 4.6   Artificial neural network (radial basis function)-based MPPT controller.

Another MPPT techniques used in PV system is Adaptive neuro-fuzzy inference system (ANFIS), which combines both fuzzy logic and neural network. The internal data training is performed using NN technique, and the outer data training is performed using fuzzy logic. This technique is mainly used for tracking the maximum point rapidly and to determine an efficient power output. Biologically inspired control algorithm is also used in literature to track the maximum point. The bio-inspired MPPT control of PV system includes swarm optimization, birds flocking, and fish schooling. In bio-inspired MPPT technique, the PV module acts as the particle and the MPP acts as target. The particles tend to find the target automatically by using these control strategies. The efficiency of the bio-inspired MPPT technique is shown to be 12.19% more than that of conventional technique (Karami et al., 2017).

4.2.2 SOLAR TRACKING

The solar tracker is a mechanism to trace the solar light and position of the PV panel toward maximum solar ray in order to obtain maximum power. This control strategy increases the total amount of energy collected by PV panel based on the area. The partial shading effects are drastically reduced with the implementation of this method (Mousazadeh et al., 2009). The addition of mechanical structure in the PV panel increases the cost of installation of PV system. Thus, the technical advancements of producing low-cost high-concentrated PV array can reduce the installation cost and provide better efficiency to the system (Rubio et al., 2007). There are three types of solar tracking methods: fixed surface-based tracking, one axis tracking system, and bi-axis tracking system. The pointing error and cost of installation are the major parameters in selecting the type of tracking technique. The fixed tracking controller provides low-cost installation and the solar panel is fixed to certain changes according to the daily pattern. Thus, no external sensors are required in this method. The major disadvantage of this system is that the panel is altered in fixed pattern, thus changes in the climatic condition are not traceable (Sefa et al., 2009). In one axis tracking method, the PV can be tilted in single direction only. The sensors are placed on the top and bottom side of the PV panel. This is mostly preferred tracking method for its efficiency and low cost than two axis orientation. Here the PV panel can track most of the solar rays by tilting according to the solar intensity (Lee et al., 2009). The bi-axis tracking method is preferred only for very high installing area where there are more passing clouds. In bi-axis method, the tracking can be done in both the ways front–back and right–left. Thus, they require more sensor and motor which increases the cost of over system (Abdallah & Nijmeh, 2004). So they are used mainly for region where the solar panels are installed in MW ranges.

Comparison of these methods is shown in Table 4.2. The fixed method uses simple mathematical formula to determine the solar movement and the microprocessor unit tilts the panel based on the fed data. The one axis and bi-axis control technique uses electro-optical unit which is installed using sensors and motors. Basically, pyrheliometer or light detecting sensor is used to estimate the solar position to feed the information to the microprocessor and perform the control action as shown in Figure 4.7 where one-axis tracking is implemented (Camacho & Berenguel, 2012).

TABLE 4.2   Comparison of Different Solar Tracking Methods for PV System.

Technique	Evaluation technique	Sensor requirement	Complexity	Cost	Efficiency	Gain achieved (fixed)
Fixed	Theoretical	No	Low	Less	Very low	0
One axis	Actual	Yes	Medium	Medium	Medium–high	10–20% (seasonal)
Bi-axis	Actual	Yes	High	Very high	High	57% (yearly)

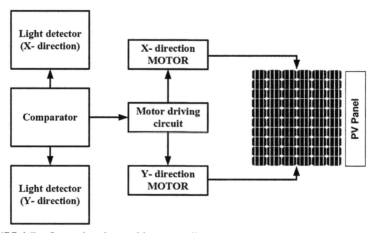

FIGURE 4.7   One-axis solar tracking controller.

4.2.3   FRT AND GRID CAPABILITY (GC)

High penetration of PV system introduces negative impact in the grid. Issues such as stability, power quality, efficiency, and reliability are becoming more important to provide full support to the grid-connected PV system (Yang et al., 2014). The FRT-based control strategy for solar-based power generation technique is used to provide better stability to the system when connected to the grid. They are also termed as LVRT or under voltage ride through (UVRT). LVRT is not implemented in small-scale PV system. The other controllers such as MPPT and solar tracking are much focused for stable operation. But due to reactive power compensation, the FRT-based control technique is implemented (Yang et al., 2016). The grid code for renewable energy-based distributed system should be maintained. The grid fault which occurs should be eradicated as soon as possible, thus the need for FRT-based control strategy is much needed mainly on large-scale PV application. Basically, the single-phase grid-connected PV system includes two control loops (Lakshmi & Hemamalini, 2016):

1) an inner current control loop to overcome the power quality issue and protect the system with high inrush current and
2) an outer voltage control loop where the desired current reference is generated to control the inner loop control using power and voltage obtained.

The stability of the whole control system depends on the accuracy and speed of voltage sag detection and synchronization of the system with the grid under faulty condition. The sag detection techniques are very essential to determine efficiency and stability of the system. The important methods that are used to increase the efficiency of sag detection technique are resonant controller, phase-locked loop (PLL) controller, root mean square (RMS) method, and peak voltage method (Aziz, 2012).

The resonant controller is the standard controller technique as PI controller. The resonant controller is based on proportional-resonant (PR) approach which is used to control the DC signal of the system. In three-phase system, the dq transformation method is used to transform the signal into the DC. The resonant control strategy can track the gain at the infinite gain at its resonance frequency, thus permitting the low pass transformation. The PLL behavior of the PV system fails during the grid fault or during the unsymmetrical behavior of grid synchronization. In PLL method, the stability of the system is affected by the sudden change in the grid condition since they are based on Park's dq transformation. The strategy is carried out into two stages. First, the PLL is stabilized within the unsymmetrical grid condition and then the current is ensured to be in limit of specified code (Marinopoulos et al., 2011).

The RMS method is based on the determining the phase voltages. The voltage between the phases is calculated over a half period. Thus, by detecting the value of each phase, the voltage sag can be determined. In peak voltage method, the absolute voltage of each period is calculated and the voltage sag is recorded over the period of time. They are similar to the peak value method of detecting the fault where the parameters are calculated for half cycle. Among all the sag detection method, RMS-based voltage sag detection method acts fast to detect the voltage dip and they are in compliance with the standard grid code (Styvaktakis et al., 2001).

The GSC-based control strategy also includes the reactive power compensation method. When voltage sag is detected, the PV system enters in LVRT mode. The voltage drop should withstand for specified duration to follow the grid code. The operation of LVRT and fault condition is mentioned in Figure 4.8, which shows the region of grid connected system and islanded mode. During the voltage sag, the control system should deliver the reactive power to grid in order to support the grid recovery. The LVRT is mainly used for large-scale applications to support the grid and protect the system. The reactive power injection into grid when LVRT is functioning can stabilize the grid voltage. The mainly used reactive power injection methods in three-phase PV applications are (Yang et al., 2014):

1) unity power factor control strategy,
2) positive and negative sequence control strategy,
3) constant active power control strategy, and
4) constant reactive power control strategy.

The presence of interaction between the voltage sequence and current sequence under grid fault can cause unbalanced grid condition. Thus, the oscillation will be present in both active and reactive power control strategy. To overcome this, zero sequence control path is introduced which increases the control freedom of the system and eliminates the oscillation, thus increasing the stability of the system. The reactive power injection in single-phase PV system is done using two control strategy:

1) peak current control strategy and
2) active current control strategy.

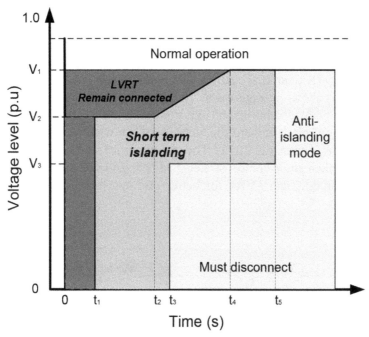

FIGURE 4.8 Compatible region of LVRT and anti-islanding operation of PV-based power generation system.

Both the control strategy has benefit of low cost and fast computational time. The LVRT strategy is further required for normal operation of PV system during grid fault condition. Thus, they are most preferred research area among researchers and industries.

4.3 CONTROL STRATEGIES IN WIND ENERGY CONVERSION SYSTEMS

Wind energy is one of the most promising renewable energy sources of energy due to its eco-friendly nature. The growing demand due to vast urbanization is met by installing large-scale wind turbines. Thus, integration of large-scale wind turbine into grid requires more sophisticated control units for better efficiency and enhanced grid synchronization. Another area of research is minimizing the production cost to give it a competitive edge over another power source (Tiwari & Babu, 2016). The development of wind energy conversion system (WECS) during recent years has been very dynamic. The electrical design and the control strategy employed in the WECS are very much differentiating from each other. Basically, the control ability of WECS is classified as speed control and power control methodology (Kumar et al., 2016). The operation regions associated with the WECS is shown in Figure 4.9. The majority of the wind turbine used today are three bladed horizontal axis configuration which provides better efficiency and robust toward the stability of the wind turbine instalment (Bertašienė & Azzopardi, 2015). The major control strategies available in WECS are pitch angle control strategy, MPPT, grid side and machine side control technique, and LVRT for better grid synchronization (Tiwari & Babu, 2016).

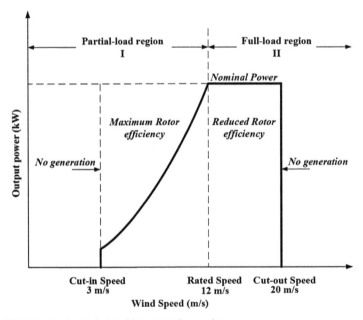

FIGURE 4.9   Typical wind turbine operating region.

4.3.1 PITCH ANGLE CONTROL

The wind turbines are classified into two categories: fixed pitch and another is variable pitch. Fixed pitch is the cheapest form of wind turbine which is employed only in small WECS due to inability of varying pitch angle to capture the maximum kinetic energy from available wind. Variable pitch control wind turbines can be operated near the optimum power efficiency but they require additional control unit to control to provide different pitch angle based on the wind speed (Munteanu et al., 2008). The pitch angle regulates the output power mechanically by varying the blade angle based on the angular speed and direction of wind. During wind speed lower than that of rated wind speed, the pitch angle is kept to maximize the rotor speed of wind turbine to extract the maximum available power from the wind. And during wind speed higher than that of rated wind speed, the pitch angle is adjusted to maintain the optimum power of the wind turbine so as to protect the system (El-Tous, 2008). They highly absorb the nonlinearity of the system, thus protecting it from the sudden wind gust. The pitch angle controller generally consists of motor and an electromechanic actuator to control the angle of blade. The pitch actuator of wind turbine can be installed in collective blade system or individually (Njiri & Soffker, 2016). The individual pitch mechanism is expensive method and it is used only in large-scale wind turbine. Collective pitch angle controller is very much popular for being cost effective and highly efficient. Pitch control system is classified into two types, hydraulic controller and electric pitch controller (Tong, 2010).

4.3.1.1   HYDRAULIC PITCH ANGLE CONTROL

The hydraulic pitch angle controller uses hydraulic actuator to control the blade of wind turbine. The hydraulic actuator is placed in the wind turbine hub along with the accumulator tank which provides the linear movement in the blades. The hydraulic pump which is situated in the nacelle of the turbine is used to generate the corresponding energy for the rotatory operation of the blade. The hydraulic pitch angle controller has a significant advantage such as low complexity, safer operation, and robust toward nonlinear characteristic of wind speed. The installation cost of hydraulic controller is not very much expensive when compared with the electro-mechanical pitch angle controller. The major disadvantage of hydraulic pitch angle controller is the frequent maintenance and possibility of oil spill in the turbine during natural calamity. The oil used in hydraulic pump need to be replaced after a specified interval of time. Thus, it increases the operational and maintenance cost of the system. Figure 4.10 shows the basic configuration of hydraulic system. In certain application, the position sensor is required to feed the data of current pitch angle degree of the blade to the hydraulic actuator (Tiwari & Babu, 2016).

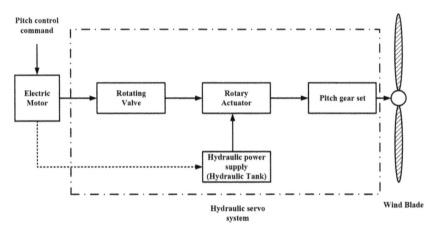

FIGURE 4.10   Basic configuration of hydraulic pitch angle controller.

4.3.1.2   ELECTRIC PITCH ANGLE CONTROL

Electric pitch angle controller constitutes of both electrical and mechanical methodology to alter the angle of the blade individually or collectively. The electromechanical pitch angle controller is equipped with an electric motor, energy storage system to run the motor, sensor to measure the wind velocity and direction, and a gear box to adjust the speed of the motor as shown in Figure 4.11 (Geng & Geng, 2014). The response time of electric pitch angle controller is faster than that of hydraulic controller. They are expensive for installation since they require a power back up when compared to the hydraulic pitch angle controller but the operation and maintenance of this controller is very minimal (Tavner et al., 2007). Various electric pitch angle controllers are used in the literature. The most preferably used electric pitch angle controllers are explained in the following paragraph.

The most basic and suitable controller for small-scale WECS is PI/PID controller. This conventional controller derives the pitch reference from the input parameters such as rotor speed, generator power, and wind speed. The wind speed-pitch angle curve is designed by the manufacturer to specify the blade angle for the corresponding wind speed (Hau, 2006).

The conventional controllers fail to track the rapid variations in the wind speed; thus, to increase the control performance of the nonlinear characteristic of the wind turbine, gain scheduling method is employed along with the conventional controllers (Knight & Peters 2005). The sensitivity of the aerodynamic torque which are the variations of the output power with respective to change in pitch angle is minimized. The reliability of the system is increased when the conventional controller is used along with the gain scheduling. The major disadvantage of conventional pitch controller is nonability to handle the nonlinearity of the system. The response time of the conventional converter is also high, thus cannot be used in the region where there are sudden changes in the wind flow. They are mostly suitable for small WECS for being cost-effective (Rugh & Shamma, 2000).

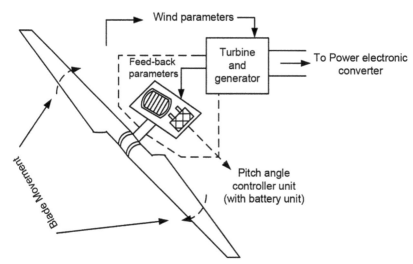

FIGURE 4.11   Basic configuration of electric pitch angle controller.

Sliding mode controller (SMC), feed-forward and feed-backward, H-infinity, and linear–quadratic–Gaussian (LQG) controls are some of the robust-based pitch angle controller used in the literature. SMC controller consists of simple design and cost-effective solution for pitch angle controller. But they highly depend on the mathematical model of the wind turbine system, thus they require previous knowledge of the system. The mechanical stress of the system is also increased due to sudden changes in control variables (Beltran et al., 2009). H-infinity-based pitch angle controller is best alternative to absorb the nonlinear parameter of the WECS. They are mostly preferable in a location where there is rapid change in wind speed. The performance of the output power is also increased. The only disadvantage is that designing the parameters for H-infinity is complex. The constraints for the parameters are intricate of the weighing functions (Laks et al., 2009). LQG-based pitch angle controller is proposed in the literature (Yao, 2009). The LQG controller provides robustness to the phase and gain margin of the system. The LQG controller fails to achieve the stability of characteristics of WECS when subjected to large distur-bance. Feed-forward and feed-backward are well-known controllers in the field of control system. They are mainly designed in order to reduce the fatigue of the load of the system and increase the turbine life. This control system mainly uses wind speed as the control input. LIDAR is used to provide wind speed data to the system. Thus, this increases the overall cost of the system and complexity in handling the LIDAR device (Muller et al., 2002). The robust-based pitch angle controller has good efficiency in terms of providing robustness to the nonlinearity of the system and compensating the uncertainties of WECS. But the control scheme is very complex when compared with other controllers, thus they are not suitable for large-scale WECS.

The most suitable and efficient pitch angle controller is based on soft computing artificial intelligence-based controllers. Artificial intelligence has a unique technique to enhance the efficiency and the response time of the system. The soft computing technique has an ability to solve a broad range of problems (Chedid et al., 1999). In general, soft computing tech-niques such as fuzzy logic controller (FLC), neural network controller, genetic algorithm (GA), and hybrid of the above controllers are majorly used. The inputs used in soft computing-based pitch angle controllers are generator speed, wind speed, aerodynamic torque, aerodynamic power, and generator power. The FLC is very simple and recent advancements in its implementation are widely preferred for control system. The FLC-based controller has an advantage of effortless in terms of designing a control unit which is adapted in wide area of research. The control parameters of FLC-based controller can be altered based on the system requirement swiftly. The FLC-based control strategy is purely based on the human knowledge about the system. The memory allocation is the major issue in the FLC for performing efficient control when the system is subjected to major climatic changes (Van et al., 2015). The artificial neural network (ANN)-based control of pitch angle of blade has the same features of FLC-based controller but they have slightly faster response than FLC-based control strategy. The ANN controller can be adopted in any varying conditions. The ANN controller is also user knowledge-based controller. The efficiency of ANN controller can be increased by providing suitable variables and regular training of data (Jafarnejadsani et al., 2013). The GA-based pitch angle controller is also used by some researchers to alter the blade angle for obtaining optimum power from available wind. GA controller is developed to stabilize the system during high nonlinearity in wind speed. They are used to estimate the reference pitch angle for different wind speed. GA technique is employed during low wind speed to maximize the power obtained from the generator (Tiwari & Babu, 2016).

Hybrid controller is used to overcome the drawback of all the above pitch angle control strategy. The hybrid controller uses two or more above control strategy to provide efficient and maximum stability to the system (Abdullah et al., 2012). The major combination of hybrid controllers is ANN and GA as shown in Figure 4.12. The FLC and ANN are also combined and used extensively named as ANFIS. Each controller in the hybrid system performs different roles based on the demand. As specified earlier, hybrid controller provides reliable solution for nonlinear system subjected to input constraints. The main disadvantage of hybrids controller is its cost and implementation, but the payback period of the system is quicker when hybrid controller is used (Lin et al., 2011).

Pitch angle controller has various advantages such as controllability in all wind speed region, controllability in individual blade based on the direction, and velocity of wind. Table 4.3 shows the comprehensive parameters of different pitch angle controller which are employed in WECS. The pitch angle controller can maximize power production during low wind speed region and during high wind speed region; the pitch angle reduces the efficiency to produce optimum rated power. The area to be focused is the response time of pitch angle controller which is slower when compared with another control techniques of WECS. The maintenance of pitch angle controller is also a major concern. The installation cost of the pitch angle controller purely depends on the location and wind availability of the system. Because of high installation and maintenance cost, they are used only in large wind power plants (Tiwari & Babu, 2016).

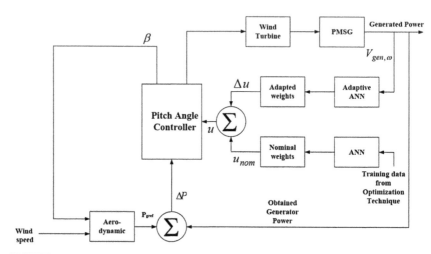

FIGURE 4.12   ANN- and GA-based hybrid pitch angle controller.

TABLE 4.3   Different Pitch Angle Control Techniques in WECS.

Technique	Reliability	Response speed	Complexity	Performance in rapid wind speed	Cost
Hydraulic	High	Slow	High	Low	High
PI/PID	Low	Slow	Low	Low	Less
Sliding mode control	Medium	Moderate	Medium–high	Low	Medium–high
Feed-forward/ feed-backward	Medium– low	Moderate	Medium– low	Low	Medium–low
H-infinity	Medium– high	Moderate– high	Medium– high	Medium–high	Medium–high
Linear– quadratic– Gaussian	Medium– high	Moderate– high	Medium– high	Medium–low	Medium–high
Fuzzy logic	High	Fast	High	High	Moderate
Neural network	High	Fast	High	High	Moderate
Hybrid	High	Very fast	Medium	Very high	Depends

4.3.2 MPPT CONTROL

MPPT algorithm is necessary for WECS to track maximum power for available wind speed. There is a specific generator speed where the maximum power is obtained beyond which the generator deteriorates and the efficiency is reduced. The MPPT controller tracks the optimum speed of the generator and extracts the maximum power from the available wind speed. The MPPT controller is generally employed, where the wind speed is between the cut-in wind speed and rated wind speed (Mahela & Shaik, 2016). When the wind speed exceeds the rated wind speed, the MPPT algorithm tends to stabilize the output power by reducing the generator efficiency, thus protecting it from surges and getting overloaded (Lin & Hong, 2010). The MPPT control strategy is generally associated with the power electronics system of the WECS as shown in Figure 4.13. MPPT control technique generates the duty cycle for the switches of converter to control the output power. There are numerous MPPT controller specified in the literature.

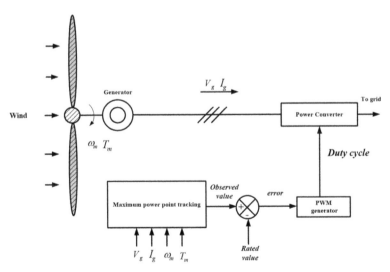

FIGURE 4.13 MPPT controller for wind energy conversion system.

The primary MPPT which were used in the WECS are power signal feedback (PSF), hill-climb search (HCS), or P&O, tip-speed ratio (TSR), and optimal torque control (OTC). The PSF control method uses DC voltage and current as the feedback for the control unit as specified in the name. PSF method requires previous knowledge of the system. Previously, PSF-based control strategy generator power and mechanical shaft speed were used for the implementation (Barakati et al., 2009). To track the shaft speed, a sensor was required, which increases the cost and size of the system. The optimal power is obtained from the relation between the parameters using the lookup table as specified by the manufacturer. The major drawback of PSF method is its complexity to implement (Lu et al., 2009). The HCS or P&O method is very basic conventional MPPT controller, and they operate similar to that of solar-based MPPT control. The HCS-based controller is very much popular in small WECS for its simplicity and low cost (Daili et al., 2015). As in solar-based control strategy, this control technique fails to track the rapid variation in the wind speed. Thus, they can only be employed where there is constant or less variable wind speed (Kumar & Chatterjee, 2016). TSR-based MPPT technique maintains the ratio between the tip of the blade and rotor speed of the generator to the optimum value in order to achieve maximum power irrespective of wind variations (Yokoyama et al., 2011). A feedback controller is required to provide an input to the control unit as the difference between the actual and the optimal value (Nasiri et al., 2014). Based on the feedback, the generator speed is altered to maintain the optimal speed of the generator (Li et al., 2015). Though the TSR-based MPPT method is simple in implementation, the operation and maintenance cost of TSR controller becomes expensive. This method also requires the precise measurement of wind speed which further increases the cost and complexity of the system (Abdullah et al., 2012). OTC-based MPPT controller alters the actual generator torque of the system based on the optimal torque of the rated speed. Each wind speed has a reference power torque where the maximum power is obtained. Thus, the actual torque is compared with the reference torque of the available wind speed and error signal is generated which is then fed to the control unit to maintain the optimal torque. The major drawback of this controller is that it does not measure the wind speed directly. Hence, a smaller wind speed variation cannot be observed in the OTC control technique in specified time interval (Nasiri et al., 2014).

The overall drawbacks of the above controller are minimized using the soft computing-based controller. Soft computing-based MPPT controller does not require any mathematical model or preknowledge of the system parameter. Soft computing-based MPPT controller has faster response time than other conventional controller since they require no sensors to measure the parameter of the system. As stated earlier, the soft computing techniques such as fuzzy logic, ANN, hybrid controller, and multivariable controls were used for WECS control. FLC-based controller has faster response time toward change in the system dynamics without estimation of the system parameter (Tripathi et al., 2015). The FLC has an ability to absorb the nonlinearity of the input variable which is very much essential in WECS. The efficiency of the FLC purely depends on the designers' knowledge about the system parameters, parametric error analysis, and rules base of the controller. The FLC-based controller generates pulse to control the power electronics converter (PEC) associated with the WECS by generating and stabilizing the output power (Tiwari & Babu, 2016). The input variable for FLC technique is usually used as output power, generator power, rotor speed, and mechanical torque. The input variable is then fed to the membership function in FLC to fuzzy the actual data, then rules are designed using the extensive knowledge about the system, and at last the defuzzification process takes place where suitable duty cycle is generated for the present wind speed (Liu et al., 2015). ANN-based MPPT controller is similar to FLC technique as they require preknowledge about the system. Instead of rules-based inference system, it uses hidden layer as the processing input variable (Cirrincione et al., 2013). The ANN controller uses DC output voltage and current, output torque, rotor speed, and wind speed as input variables. The output of the ANN control technique is reference torque, reference power, and rotor speed. The speed of convergence of maximum point from the operating point purely depends on the weights allotted in each layer (Belmokhtar et al., 2014). Hybrid controller-based MPPT technique is best suitable controller to optimize and track the efficient power without compromising the stability of the system. Many conventional and soft computing techniques are combined to track the maximum power from the wind (Assareh & Biglari et al., 2015). They have advantage of faster response time, lower risk of wear–tear of the system, increase life span of generator and turbines, and faster payback period by generating maximum power, thus increasing the yield as described in Table 4.4. The major drawback of hybrid system is its cost and complexity of designing the system (Yin et al., 2015).

TABLE 4.4 Comparative Analysis of Different MPPT Controller Used in WECS.

Techniques/parameter	PSF	HCS	TSR	OTC	SCT	Hybrid
Complexity	Simple	Simple	Simple	Simple	Complex	Complex
Wind speed measurement	Required	Not required	Required	Not required	Not required	Not required
Convergence speed	Fast	Low	Fast	Fast	Medium	Fast
Sensitivity	Yes	No	No	Yes	No	Depends
Tolerance to rapid variation	Moderate	Low	Moderate–high	Moderate–high	High	Very high
Memory requirement	Required	Not required	Not required	Depends	Not required	Depends
Prior knowledge	Required	Not required	Required	Required	Not required	Depends
Efficiency	Moderate	Low	Very high	Moderate–high	High	Very high

4.3.3 GRID SIDE AND MACHINE SIDE CONTROLLER

The machine side controller (MSC) controls the speed of the WECS system to capture the maximum power. MSC changes the speed of the rotor to the optimum value to enhance the output power as well as stability of the system. The grid side controller is utilized to control the grid parameters such as active and reactive power. The GSC observes the DC link voltage of the system which provides suitable reference of the changes occurring in the system (Jain et al., 2015). The typical configuration of GSC and MSC is shown in Figure 4.14.

MSC are basically having two control strategies to control the rotor speed. They are direct torque control (DTC) and field-oriented control (FOC). Both the controllers have similar characteristic and performance in dynamic condition (Merzoug & Naceri, 2008). DTC controls the torque and power directly. Since direct controller is employed, it has faster response and less complexity (Taib et al., 2014). The DTC has only one outer loop control, where the hysteresis compensator and flux angle is directly used to generate the switching pulse for the PEC. The transformation of reference frames between the control loop is also eliminated which reduces the complexity of the system. The ripple in torque and current determines the performance of the system. Since the measurement of speed is not required in DTC system, the need for rotor speed sensor is eliminated. The DTC-based controller has a faster response when compared with other control methods (Bowes et al., 2001).

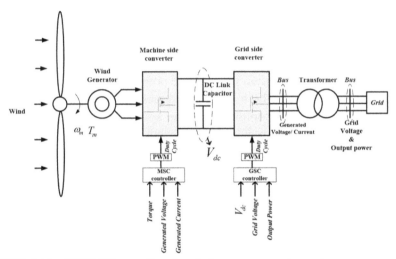

FIGURE 4.14   GSC/MSC controllers for wind energy conversion system.

FOC-based control techniques have dual control loop strategy. The outer loop control requires rotor position and speed to generate the reference current for all the three phases of the system. The inner loop is based on synchronous reference frame (Levi, 2011). The maximum electromagnetic torque is obtained by setting the d-axis current to zero (Freire et al., 2012). The developed electromagnetic torque is controlled using q-axis. FOC controls the current parameter of the system directly which increases the overall efficiency by utilizing the line current for the production of torque (Emna et al., 2013).

Grid side controller (GSC) is independent of the type of generator or converter used in the system. They focus mainly on the efficient and stable grid integration of the system. Voltage-oriented controller (VOC) and direct power controller (DPC) are the two types of GSC methodology which are widely used in wind energy system (Li et al., 2012). VOC is similar to FOC, it also has dual control loop. The DC link voltage or outer loop control and inner current control loop are two control loops present in VOC. The unity power factor of the system can be achieved when the current in q-axis is set as zero (Brekken & Mohan, 2007). The VOC strategy has high steady performance and faster response as they measure DC link voltage directly. The power quality of the system is also improved which is a primary factor in grid integration. The only disadvantage of VOC is the stability of active and reactive components of the system and requirement of reference frame (Dai et al., 2009).

DPC-based control strategy also consists of two control variables, and active and reactive powers. The inner current control loop in DPC is eliminated and they do not require any PWM technique. The unity power factor can be achieved when the reactive power is set to zero. Since only one control loop exists, there is no requirement of coordinate transformation which reduces the complexity of the system (Dannehl et al., 2009). The computation time is reduced and they have faster dynamics (Noguchi et al., 1998). They have very high robustness toward the uncertainties in WECS. The only disadvantage in DPC method is the requirement of filter inductance and sampling frequency for variable switching frequency which increases the cost of the control unit (Zhao et al., 2013). The need of filter inductance also increases the THD of the system with high current ripple. Table 4.5 shows the detailed analysis of MSC/GSC control strategy used in WECS. From the analysis of the literature, VOC-based grid side controller and FOC-based MSC are best suitable for grid integration of WECS to enhance the performance and efficiency of the system.

TABLE 4.5   GSC-/MSC-based Technique for WECS.

Parameter	MSC		GSC	
	FOC	DTC	VOC	DPC
Dynamic response time	High	Low	High	Low
Implementation	Complex	Simple	Complex	Simple
Coordinate transformation	Required	Not required	Required	Not required
Power quality	Better	Poor	Better	Poor
Internal current regulation loop	Required	Not required	Required	Not required
Power and current ripple	Less	More	Less	More
Parameter sensitivity	Sensitive	Insensitive	Sensitive	Insensitive
Power quality	Better	Poor	Better	Poor
DC link voltage ripple	–	–	Low	High
Torque ripple	Less	More	–	–
Rotor position sensor requirement	Required	Not required	–	–

4.3.4 LOW VOLTAGE RIDE THROUGH

LVRT is one of the most preferred power quality solutions to meet the grid requirements in WECS. The grid voltage dips, synchronization mismatch between generated active power, and active power delivered to the grid are the major concerns where LVRT is focused. The requirement of grid code specified for wind energy should be met for proper and efficient interconnection of generator to the grid. The LVRT control makes the system connected to the grid even during fault occurrence. The LVRT delivers the reactive power to the grid in order to maintain the grid voltage constant. The major concern which occurs in the grid is the voltage sag which not only affects the power quality of the system but also damages the generator and PEC of WECS by constantly heating it up. LVRT is supported by the wind generator when the system is subjected to fault and the grid voltage is reduced. The sudden large change in load also causes the voltage dip in the grid. Thus, to overcome this issue and to maintain the grid stability while reducing the risk of voltage collapse, LVRT transfers the reactive power. Initially, the grid codes were designed only for synchronous-based wind turbine generator. But later, the need of uniform grid code to reduce the risk of power outage and frequent disconnection of WECS from grid, Indian Wind Grid Code (IWGC) specified certain constraints for large producers of power using wind energy. The operating region of wind farms connected to 66 kV grid is shown in Figure 4.15. The time of fault and disconnection of WECS from grid is mentioned in the grid code for maintaining the stability of the grid (Singh & Singh, 2009).

The lower limit of voltage is taken as 15% of the nominal voltage V_f. The V_{pf} is specified by the IWGC which is the minimum voltage requirement for 66 kV grid. The roles of LVRT when the grid fault occurs or voltage dip occurs are:

- to remain connected to grid until the grid voltage reaches below the limit specified and
- to inject the reactive power during grid fault.

The fault clearing time which is specified for different nominal voltage level specified by IWGC is tabulated in Table 4.6 (Mali et al., 2014).

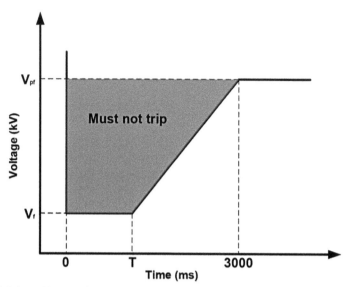

FIGURE 4.15   Characteristic of fault ride through.

The parameters such as fault severity, location of fault, and the stability condition of the system the factors such as voltage sag, voltage swell, or temporary outrage can happen. The duration and the magnitude of the fault purely depend on the control strategy used and the fault character-istic. The recovery from the fault depends of the reactive power support of the system and the strength of the grid synchronization.

TABLE 4.6   Fault Clearing Time for Various System Nominal Voltage Levels.

Nominal system voltage (kV)	Fault clearing time (ms)	V_{pf} (kV)	V_f (kV)
400	100	360	60
220	160	200	33
132	160	120	19.8
110	160	96.25	16.5
66	300	60	9.9

Grid code specified by the IWGC typically is subjected to only large scale wind farms. The small WECS connected to the distribution network are exempted and have very minimal code to follow. The grid code stipu-lates the wind farm to follow the strategy of conventional power plant for power system control such as voltage and frequency requirement. The wind farm grid code has increased limits for frequency and voltage constraints. The most commonly used control strategy includes frequency control, active power regulation, reactive power management, and the power factor (Tsili & Papathanassiou, 2009).

The major contribution of LVRT remains to withstand the voltage sag for specific duration until the prescribed minimum voltage. The wind turbine generator should be connected to the grid for certain time during the fault and ensure that there is no generation loss during the period. The regular disconnection of WECS from the grid makes a negative impact when large-scale wind farms are employed. Thus, LVRT is a minimum and essential requirement for any large-scale wind farm to gain the immunity against self-clearing faulty conditions (Benbouzid et al., 2015).

4.4 CONTROL STRATEGIES IN HRE CONVERSION SYSTEMS

Hybrid energy systems require well-organized control strategies in order to handle the heterogeneous energy (solar, wind, diesel generators, storage systems, etc.) sources that are interfaced through the converters. In particular, renewable energy sources that are intermittent in nature will cause a lot of issues when they are connected in hybrid energy system. The major issues are stability in electric network and power quality. It is a very important task for the energy managers to handle these uncertainties caused by renewable sources. Hence, there should be perfect supervisory control scheme which will take care of voltage and frequency profiles. There are different control strategies available in the literature which are suitable for integration of different energy sources and their energy scheduling.

In a broad sense, the control strategies are classified as:

1) centralized control,
2) distributed control,
3) hybrid control, and
4) multiple control system.

In all four categories, each resource has its own number of local controller (slave controller) and centralized controller (master controller) that determines the optimal operation of the source based on the available current information. At first stage, all energy sources and demand are forecasted and at later stage energy sources, demand and scheduling of energy sources and storage devices are optimized to achieve optimal energy flow in hybrid system. An intelligent energy flow management in hybrid energy system is shown in Figure 4.16 (Chauhan & Saini, 2014).

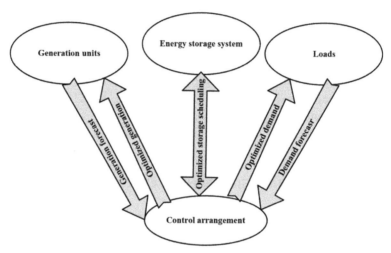

FIGURE 4.16   Intelligent energy flow management in hybrid energy system.

4.4.1   CENTRALIZED CONTROL

In this control arrangement, the whole system will be monitored and controlled by one master controller (centralized controller) and several local controllers (slave controllers) for different RESs, storage systems, and diesel generations. The central controller will gather the data from all local controllers and remote terminals in order to take necessary actions required to maintain the stability. The measured data from all the meters located at different zones will be sent to central controller and the control or acknowledge information will be sent to specified controller. The control and data flow are shown in the Figure 4.17. The central controller acts as supervisor for energy management and makes the decisions on control actions on the data retrieved and energy flow constraints. Depending upon the availability of resources of power generation and load forecasting data, it will decide the energy flow from various energy resources in the integrated system. This kind of control scheme is well suited for multiobjective energy flow management that leads to global optimum values. This control scheme has disadvantage that it requires heavy computation time and may lead to failures with single point inaccuracy.

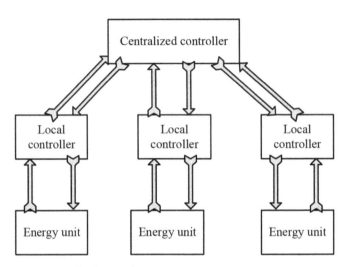

FIGURE 4.17   Centralized control.

4.4.2   DISTRIBUTED CONTROL

In this control scheme, each energy source will have a local controller and the data from all nearby generations will be sent to this controller in order to support the decisions and control actions. The data and control flow of this control scheme are shown in the Figure 4.18. In distributed control, the master control action like discussed in the above scheme disappears and this globalized control for whole network will be accomplished by sharing the data among neighboring controllers (local controllers). In this scheme, the computational time is reduced as all the local controllers will share the task of data processing and generates the required control action needed. But, the controllers which are adjacently and vertically communicated cause the complexity in the communication configuration. The application of soft computing techniques in this distributed scheme helps in reducing the network complexity and energy management of integrated system. Failure probability of this scheme is limited to local controllers but do not affect the whole energy system.

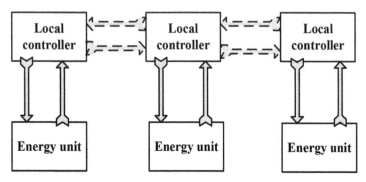

FIGURE 4.18   Distributed control.

4.4.3   HYBRID CONTROL

Hybrid control scheme is a combination of both centralized and decentralized control strategies. The data and control flow of this control scheme are shown in the Figure 4.19. In this control scheme, the renewable energy sources are combined within integrated energy network. The coordination between the central and distributed control centers will lead to global optimum. The control and monitoring tasks will be shared and executed in accordance with globalize constraints. This hybrid scheme is more suitable for heterogeneous grid than the both central and decentralized control schemes. It also needs less computational time and hence the data handling and processing will no more be an issue.

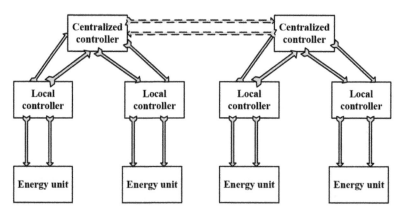

FIGURE 4.19   Hybrid control.

4.4.4 MULTIPLE CONTROL

The data and control flow of multilevel control arrangement is shown in the Figure 4.20. This is designed by adding strategic (supervisory) control to the hybrid control arrangement. At the operational level, basic decisions related to real-time operation are made, and actual control of each energy unit is performed based on the control objective of the unit very rapidly. The tactical level aims to make operational decisions for a group of local control units or the entire subsystem, with a relatively higher time frame. Strategic decisions concerning the overall operation of the system (start-up or shutdown) are made at the top level. Two-way communication exists among the different levels to execute decisions (Nehrir et al., 2011).

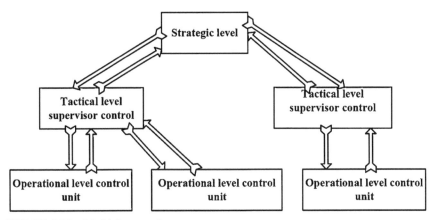

FIGURE 4.20   Multiple control.

The summary of different control strategies in hybrid system are listed in Table 4.7.

4.5   CONCLUSION

Renewable-energy-based generation is gaining interest all over the world. Control strategies for renewable-based generation are given more importance to harvest more efficient energy from the source such as solar and wind. The intermittent nature of renewable energy assigns a challenging task to the control strategy to obtain a high quality and reliable power

TABLE 4.7 Summary of Hybrid Control Strategies.

Control arrangement	Summary	Advantages	Disadvantage
Centralized control	Centralized controller receives the data from the all the sources. Global optimization is achieved by the multiobjective energy management system.	Whole system is monitored and controlled by the one centralized (master) controller	Heavy computation burden Reliability
Distributed control	The measurement signals of the energy sources of the hybrid system are sent to their corresponding local controller.	Easy of plug-and-play operation. Less computation burden	Failure in the communication link makes the complexity in the communication system
Hybrid control	Centralized controller achieved the local optimization in each group and global coordination is achieved by the distributed controller among the groups.	Less computational time	Failure in the communication link makes the complexity in the communication system.
Multiple control	Centralized controller achieved the local optimization in each group and global coordination is achieved by the distributed controller among the groups.	Each controller computational burden is reduced Overcomes the single-point failure problems	Failure in the communication link makes the complexity in the communication system

source. Due to high penetration of renewable-based power generation in to the electrical grid, it is necessary to produce an efficient and stable power which follows the grid code.

In this chapter, several aspects of control strategy which are employed in solar and wind energy for stand-alone and grid-connected system were discussed. The main issue remains in efficiency, power quality, stability of system, and grid synchronization while feeding into grid. The smart technique, unconventional idea, and the performance of each control techniques are explained. Application of these techniques in future can lead to reliable generation of power with low cost and computation response speed of the system.

The control strategy which is employed where the renewable-based energy system is combined to form hybrid energy system is also presented in this chapter. The complexity of the techniques, it merits and demerits, and constraints limit of each control technique are precisely discussed in this chapter. Thus, this chapter provides complete details of control strategies present in renewable-based generation in order to provide stable and efficient power.

KEYWORDS

- **control techniques**
- **PV-based power generation system**
- **wind energy conversion system**
- **hybrid system**
- **conventional strategy**
- **soft computing intelligent control**
- **grid integration**

REFERENCES

Abbasi, T.; Premalatha, M.; Abbasi, S. A.. The Return to Renewables: Will It Help in Global Warming Control. *Renew. Sustain. Energy Rev.* **2011**, *15*(1), 891–894.

Abdallah, S.; Nijmeh, S. Two Axes Sun Tracking System with PLC Control. *Energy Convers. Manag.* **2004**, *45*(11), 1931–1939.

Abdullah, M. A.; Yatim, A. H.; Tan, C. W.; Saidur, R. A Review of Maximum Power Point Tracking Algorithms for Wind Energy Systems. *Renew. Sustain. Energy Rev.* **2012**, *16*(5), 3220–3227.

Abdullah, M. A.; Yatim, A. H.; Tan, C. W.; Samosir, A. S. In *Particle Swarm Optimization-based Maximum Power Point Tracking Algorithm for Wind Energy Conversion System.* Proceedings of the International Conference on Power and Energy, 2012, 65–70.

Assareh, E.; Biglari, M. A Novel Approach to Capture the Maximum Power from Variable Speed Wind Turbines Using PI Controller, RBF Neural Network and GSA Evolutionary Algorithm. *Renew. Sustain. Energy Rev.* **2015**, *51*, 1023–1037.

Aziz, A. Modeling and Simulation of Dynamic Voltage Restorer in Power System. Thesis, Master of Science, Azhar University, 2012.

Barakati, S. M.; Kazerani, M.; Aplevich, J. D. Maximum Power Tracking Control for a Wind Turbine System Including a Matrix Converter. *IEEE Trans. Energy Convers.* **2009**, *24*(3), 705–713.

Bazilian, M., et al. Re-considering the Economics of Photovoltaic Power. *Renew. Energy.* **2013**, *53*, 329–338.

Belmokhtar, K.; Doumbia, M. L.; Agbossou, K. Novel Fuzzy Logic Based Sensorless Maximum Power Point Tracking Strategy for Wind Turbine Systems Driven DFIG (Doubly-fed Induction Generator). *Energy* **2014**, *76*, 679–693.

Beltran, B.; Ahmed-Ali, T.; Benbouzid, M. E. High-order Sliding-mode Control of Variable-speed Wind Turbines. *IEEE Trans. Ind. Electron.* **2009**, *56*(9), 3314–3321.

Benbouzid, M.; Muyeen, S.; Khoucha, F. An Up-to-date Review of Low-voltage Ride-through Techniques for Doubly-fed Induction Generator-based Wind Turbines. *Int. J. Energy Convers.* **2015**, *3*(1), 1–9.

Bertašienė, A.; Azzopardi, B. Synergies of Wind Turbine Control Techniques. *Renew. Sustain. Energy Rev.* **2015**, *45*, 336–342.

Bowes, S. R.; Grewal, S.; Holliday, D. Novel Adaptive Hysteresis Band Modulation Strategy for Three-phase Inverters. *IEEE Proc. Electric Power Appl.* **2001**, *148*(1), 51–61.

Branker, K.; Pathak, M. J.; Pearce, J. M. A Review of Solar Photovoltaic Levelized Cost of Electricity. *Renew. Sustain. Energy Rev.* **2011**, *15*(9), 4470–4482.

Brekken, T. K.; Mohan, N. Control of a Doubly Fed Induction Wind Generator Under Unbalanced Grid Voltage Conditions. *IEEE Trans. Energy Convers.* **2007**, *22*, 129–135.

Camacho, E. F.; Berenguel, M. Control of Solar Energy Systems. *IFAC Proc.* **2012**, *45*(15), 848–855.

Chauhan, A.; Saini, R. P. A Review on Integrated Renewable Energy System Based Power Generation for Stand Alone Applications: Configuration, Storage Options, Sizing Methodologies and Control. *Renew. Sustain. Energy Rev.* **2014**, *38*, 99–120.

Chedid, R.; Mrad, F.; Basma, M. Intelligent Control of a Class of Wind Energy Conversion Systems. *IEEE Trans. Energy Convers.* **1999**, *14*(4), 1597–1604.

Cirrincione, M.; Pucci, M.; Vitale, G. Neural MPPT of Variable-pitch Wind Generators with Induction Machines in a Wide Wind Speed Range. *IEEE Trans. Ind. Appl.* **2013**, *49*(2), 942–953.

Dai, J.; Xu, D. D.; Wu, B. A Novel Control Scheme for Current-source-converter-based PMSG Wind Energy Conversion Systems. *IEEE Trans. Power Electron.* **2009**, *24*, 963–972.

Daili, Y.; Gaubert, J. P.; Rahmani, L. Implementation of a New Maximum Power Point Tracking Control Strategy for Small Wind Energy Conversion Systems Without Mechanical Sensors. *Energy Convers. Manag.* **2015**, *97*, 298–306.

Dannehl, J.; Wessels, C.; Fuchs, F. W. Limitations of Voltage-oriented PI Current Control of Grid-connected PWM Rectifiers with Filters. *IEEE Trans. Ind. Electron.* **2009**, *56*, 380–388.

El-Tous, Y. Pitch Angle Control of Variable Speed Wind Turbine. *Am. J. Eng. Appl. Sci.* **2008**, *1*(2), 118–120.

Eltigani, D.; Masri, S. Challenges of Integrating Renewable Energy Sources to Smart Grids: A Review. *Renew. Sustain. Energy Rev.* **2015**, *52*, 770–780.

Emna, M. E.; Adel, K.; Mimouni, M. F. The Wind Energy Conversion System Using PMSG Controlled by Vector Control and SMC Strategies. *Int. J. Renew. Energy Res.* **2013**, *3*(1), 41–50.

Freire, N.; Estima, J.; Cardoso, A. A Comparative Analysis of PMSG Drives Based on Vector Control and Direct Control Techniques for Wind Turbine Applications. *Przegląd Elektrotechniczny.* **2012**, *88*(1a), 184–187.

Geng, H.; Geng, Y. Linear and Nonlinear Schemes Applied to Pitch Control of Wind Turbines. *Sci. World J.* **2014**, 1–9.

Hau, E. Wind Turbines: Fundamentals, Technologies, Application, Economics. *Springer* **2006**, 408–411.

Ishaque, K.; Salam, Z. A Review of Maximum Power Point Tracking Techniques of PV System for Uniform Insolation and Partial Shading Condition. *Renew. Sustain. Energy Rev.* **2013**, *31*(19), 475–488.

Jafarnejadsani, H.; Pieper, J.; Ehlers, J. Adaptive Control of a Variable-speed Variable-pitch Wind Turbine Using Radial-basis Function Neural Network. *IEEE Trans. Control Syst. Tech.* **2013**, *21*(6), 2264–2272.

Jain, S.; Agarwal, V. Comparison of the Performance of Maximum Power Point Tracking Schemes Applied to Single-stage grid-connected Photovoltaic Systems. *IET Electric Power Appl.* **2007**, *1*(5), 753–762.

Jain, B.; Jain, S.; Nema, R. K. Control Strategies of Grid Interfaced Wind Energy Conversion System: An Overview. *Renew. Sustain. Energy Rev.* **2015**, *47*, 983–996.

Karami, N.; Moubayed, N.; Outbib, R. General Review and Classification of Different MPPT Techniques. *Renew. Sustain. Energy Rev.* **2017**, *68*, 1–8.

Knight, A. M.; Peters, G. E. Simple Wind Energy Controller for an Expanded Operating Range. *IEEE Trans. Energy Convers.* **2005**, *20*(2), 459–466.

Kumar, D.; Chatterjee, K. A Review of Conventional and Advanced MPPT Algorithms for Wind Energy Systems. *Renew. Sustain. Energy Rev.* **2016**, *55*, 957–970.

Kumar, Y., et al. Wind Energy: Trends and Enabling Technologies. *Renew. Sustain. Energy Rev.* **2016**, *53*, 209–224.

Kwan, T. H.; Wu, X. Maximum Power Point Tracking Using a Variable Antecedent Fuzzy Logic Controller. *Solar Energy* **2016**, *137*, 189–200.

Laks, J. H.; Pao, L. Y.; Wright, A. In *Combined Feedforward/Feedback Control of Wind Turbines to Reduce Blade Flap Bending Moments.* Proceedings of AIAA Aerospace Sciences Meeting, 2009, 687–702.

Lakshmi, M.; Hemamalini, S. Decoupled Control of Grid Connected Photovoltaic System Using Fractional Order Controller. *Ain Shams Eng. J.* **2016** (In press).

Lee, C. Y.; Chou, P. C.; Chiang, C. M.; Lin, C. F. Sun Tracking Systems: A Review. *Sensors* **2009**, *9*(5), 3875–3890.

Levi, E. *FOC: Field Oriented Control, the Industrial Electronics Handbook*; CRC Press, Boca Raton, 2011.

Li, S.; Haskew, T.; Swatloski, R. P.; Gathings, W. Optimal and Direct-current Vector Control of Direct-driven PMSG Wind Turbines. *IEEE Trans. Power Electron.* **2012**, *27*, 2325–2337.

Li, D. Y.; Song, Y. D.; Gan, Z. X.; Cai, W. C. Fault-tolerant Optimal Tip-speed-ratio Tracking Control of Wind Turbines Subject to Actuation Failures. *IEEE Trans. Ind. Electron.* **2015**, *62*(12), 7513–7523.

Lin, W. M.; Hong, C. M. Intelligent Approach to Maximum Power Point Tracking Control Strategy for Variable-speed Wind Turbine Generation System. *Energy* **2010**, *35*(6), 2440–2447.

Lin, W. M., et al. Hybrid Intelligent Control of PMSG Wind Generation System Using Pitch Angle Control with RBFN. *Energy Convers. Manag.* **2011**, *52*(2), 1244–1251.

Liu, J., et al. A Novel MPPT Method for Enhancing Energy Conversion Efficiency Taking Power Smoothing into Account. *Energy Convers. Manag.* **2015**, *101*, 738–748.

Lu, B., et al. In *A Review of Recent Advances in Wind Turbine Condition Monitoring and Fault Diagnosis. Power Electron. Mac. Wind Appl. IEEE.* 2009, 1–7.

Mahela, O. P.; Shaik, A. G. Comprehensive Overview of Grid Interfaced Wind Energy Generation Systems. *Renew. Sustain. Energy Rev.* **2016**, *57*, 260–281.

Mali, S.; James, S.; Tank, I. Improving Low Voltage Ride-through Capabilities for Grid Connected Wind Turbine Generator. *Energy Procedia* **2014**, *54*, 530–540.

Marinopoulos, A., et al. Grid Integration Aspects of Large Solar PV Installations: LVRT Capability and Reactive Power/Voltage Support Requirements. *In PowerTech. IEEE Trondheim.* **2011**, 1–8.

Merzoug, M. S.; Naceri, F. Comparison of Field-oriented Control and Direct Torque Control for Permanent Magnet Synchronous Motor (PMSM). *In P. W. Sci. Eng. Tech.* **2008**, *35*, 299–304.

Messalti, S.; Harrag, A.; Loukriz, A. A New Variable Step Size Neural Networks MPPT Controller: Review, Simulation and Hardware Implementation. *Renew. Sustain. Energy Rev.* **2017**, *68*, 221–233.

Moretti, M., et al. A Systematic Review of Environmental and Economic Impacts of Smart Grids. *Renew. Sustain. Energy Rev.* **2016**, *68*, 888–898.

Mousazadeh, H.; Keyhani, A.; Javadi, A.; Mobli, H.; Abrinia, K.; Sharifi, A. A Review of Principle and Sun-tracking Methods for Maximizing Solar Systems Output. *Renew. Sustain. Energy Rev.* **2009**, *13*(8), 1800–1818.

Muller, S.; Deicke, M.; Doncker, R. W. D. Doubly Fed Induction Generator Systems for Wind Turbines. *IEEE Ind. Appl. Mag.* **2002**, *8*(3), 26–33.

Munteanu, I., et al. *Optimal Control of Wind Energy Systems: Towards a Global Approach*; Springer Science & Business Media, 2008.

Nasiri, M.; Milimonfared, J.; Fathi, S. H. Modeling, Analysis and Comparison of TSR and OTC Methods for MPPT and Power Smoothing in Permanent Magnet Synchronous Generator-based Wind Turbines. *Energy Convers. Manag.* **2014**, *86*, 892–900.

Nehrir, M. H., et al. A Review of Hybrid Renewable/Alternative Energy System for Electric Power Generation: Configuration, Control, and Application. *IEEE Trans. Sustain. Energy.* **2011,** *2*(4), 392–403.

Njiri, J. G.; Söffker, D. State-of-the-art in Wind Turbine Control: Trends and Challenges. *Renew. Sustain. Energy Rev.* **2016,** *60,* 377–393.

Noguchi, T.; Tomiki, H.; Kondo, S.; Takahashi, I. Direct Power Control of PWM Converter Without Power-source Voltage Sensors. *IEEE Trans. Ind. Appl.* **1998,** *34,* 473–479.

Patcharaprakiti, N.; Premrudeepreechacharn, S.; Sriuthaisiriwong, Y. Maximum Power Point Tracking Using Adaptive Fuzzy Logic Control for Grid-connected Photovoltaic System. *Renew. Energy* **2005,** *30*(11), 1771–1788.

Reisi, A. R.; Moradi, M. H.; Jamasb, S. Classification and Comparison of Maximum Power Point Tracking Techniques for Photovoltaic System: A Review. *Renew. Sustain. Energy Rev.* **2013,** *19,* 433–443.

Robert, G.; Matthew, L.; Ausilio, B. Progress in Renewable Energy. *Environ. Int.* **2003,** *29*(1), 105–122.

Rubio, F. R., et al. Application of New Control Strategy for Sun Tracking. *Energy Convers. Manag.* **2007,** *48*(7), 2174–2184.

Rugh, W. J.; Shamma, J. S. Research on Gain Scheduling. *Automatica* **2000,** *36*(10), 1401–1425.

Safari, A.; Mekhilef, S. Simulation and Hardware Implementation of Incremental Conductance MPPT with Direct Control Method Using Cuk Converter. *IEEE Trans. Ind. Electron.* **2011,** *58*(4), 1154–1161.

Saravanan, S.; Babu, N. R. Maximum Power Point Tracking Algorithms for Photovoltaic System—A Review. *Renew. Sustain. Energy Rev.* **2016,** *57,* 192–204.

Sefa, İ.; Demirtas, M.; Çolak, I. Application of One-axis Sun Tracking System. *Energy Convers. Manag.* **2009,** *50*(11), 2709–2718.

Singh, G. K. Solar Power Generation by PV Technology: A Review. *Energy* **2013,** *53,* 1–3.

Singh, B.; Singh, S. N. Wind Power Interconnection into the Power System: A Review of Grid Code Requirements. *Electricity J.* **2009,** *22*(5), 54–63.

Styvaktakis, E.; Bollen, M. H.; Gu, I. Y. Expert System for Voltage Dip Classification and Analysis. *In Power Eng. Society Summer Meet. IEEE* **2001,** *1,* 671–676.

Subudhi, B.; Pradhan, R. A Comparative Study on Maximum Power Point Tracking Techniques for Photovoltaic Power Systems. *IEEE Trans. Sustain. Energy.* **2013,** *4*(1), 89–98.

Taïb, N.; Metidji, B.; Rekioua, T. A Fixed Switching Frequency Direct Torque Control Strategy for Induction Motor Drives Using Indirect Matrix Converter. *Arabian J. Sci. Eng.* **2014,** *39*(3), 2001–2011.

Tavner, P. J.; Xiang, J.; Spinato, F. Reliability Analysis for Wind Turbines. *Wind Energy* **2007,** *10*(1), 1–18.

Tiwari, R.; Babu, N. R. Fuzzy Logic Based MPPT for Permanent Magnet Synchronous Generator in Wind Energy Conversion System. *IFAC-PapersOnLine* **2016,** *49*(1), 462–467.

Tiwari, R.; Babu, N. R. Recent Developments of Control Strategies for Wind Energy Conversion System. *Renew. Sustain. Energy Rev.* **2016,** *66,* 268–285.

Tiwari, R., et al. In *Design and Development of a High Step-up DC–DC Converter for Non-conventional Energy Applications,* Proceeding PESTSE IEEE, 2016, pp 1–4.

Tong, W. *Wind Power Generation and Wind Turbine Design,* Wit Press, 2010.

Tripathi, S. M.; Tiwari, A. N.; Singh, D. Grid-integrated Permanent Magnet Synchronous Generator Based Wind Energy Conversion Systems: A Technology Review. *Renew. Sustain. Energy Rev.* **2015,** *51,* 1288–1305.

Tsili, M.; Papathanassiou, S. A Review of Grid Code Technical Requirements for Wind Farms. *IET Renew. Power Gen.* **2009,** *3*(3), 308–332.

Van, T. L.; Nguyen, T. H.; Lee, D. C. Advanced Pitch Angle Control Based on Fuzzy Logic for Variable-speed Wind Turbine Systems. *IEEE Trans. Energy Convers.* **2015,** *30*(2), 578–587.

Yang, Y.; Blaabjerg, F.; Wang, H. Low-voltage Ride-through of Single-phase Transformerless Photovoltaic Inverters. *IEEE Trans. Ind. Appl.* **2014,** *50*(3), 1942–1952.

Yang, L., et al. Coordinated-control Strategy of Photovoltaic Converters and Static Synchronous Compensators for Power System Fault Ride-through. *Electric P. Comp. Syst.* **2016,** *44*(15), 1683–1692.

Yao, X. In *LQG Controller for a Variable Speed Pitch Regulated Wind Turbine,* Proceedings Int. Conf. Intel. Human-Machine Syst. Cybernetics IEEE, 2009, pp 210–213.

Yin, X. X., et al. Sliding Mode Voltage Control Strategy for Capturing Maximum Wind Energy Based on Fuzzy Logic Control. *Int. J. Electrical P.Energy Syst.* **2015,** *70,* 45–51.

Yokoyama, H.; Tatsuta, F.; Nishikata, S. In *Tip Speed Ratio Control of Wind Turbine Generating System Connected in Series.* Int. Conf. Electrical Mach. Syst. 2011.

Zhao, Y.; Wei, C.; Zhang, Z.; Qiao, W. A Review on Position/Speed Sensorless Control for Permanent-magnet Synchronous Machine-based Wind Energy Conversion Systems. *IEEE J. Emerg. Sel. Top. Power Electron.* **2013,** 1, 203–216.

CHAPTER 5

DC–DC CONVERTERS FOR RENEWABLE ENERGY APPLICATIONS

S. SARAVANAN[1], N. PRABAHARAN[2], and N. RAMESH BABU[3,*]

[1]*Department of EEE, Sri Krishna College of Technology, Coimbatore, Tamil Nadu, India*

[2]*Department of EEE, Madanapalle Institute of Technology & Science, Madanapalle, Andhra Pradesh, India*

[3]*M. Kumarasamy College of Engineering, Karur, Tamil Nadu, India*

Corresponding author. E-mail: nrameshme@gmail.com

CONTENTS

ABSTRACT

This chapter classifies various high static gain DC–DC converter for renewable energy sources. Most of the renewable energy resources are low voltage characteristics in nature. To step up the voltage for required voltage condition, high step-up DC–DC converters are utilized. Various DC–DC converters like nonisolated, isolated, bidirectional and three port-based converters along with their types are explained in detail.

5.1 INTRODUCTION

Nowadays, the increased power demand causes the cost hike of fossil fuel and coal, and use of these raw materials for power production increases the global warming. Owing to this issue, the focus of researchers and power generating industries turned toward renewable energy sources based power generation. The renewable energy sources such as photovoltaic (PV), fuel cell (FC), and wind energy are going to guide the power demand in forthcoming years (Saravanan and Babu, 2016).

The renewable sources are generated commonly under the distributed generation in both stand-alone and grid-connected applications. The autonomous controller in distribution generation-based DC microgrid is used (Ito et al., 2004) to smother the circulating current during power generation from renewable energy sources. The microgrid with multi-layer control and smart grid communication are proposed by Wang et al.(2012)to reduce grid peak consumption and provide power balancing in generating station. In a study by Veneri et al. (2016), power structural design is proposed, which is used as a charging station for electric vehicles and also to integrate the microgrid with the support of the electric vehicle. In a study by Sun et al. (2017), microgrid is proposed to operate in both stand-alone and grid-connected modes by using smooth active synchronization controller. The benefit of this controller that it increases the idleness and flexibility and reduces the communication costs.

Unfortunately, the characteristics of renewable sources are having low-voltage characteristic in nature and are not capable of direct operation in the grid. Commonly, the PV modules are connected in series in order to reach high-voltage values and parallel for high current (Villalva et al., 2009). This requires large physical area and number of PV modules. To conquer and to convert low voltage to high voltage from renewable source, proficient DC–DC converter is needed to make better utilization and conversion. The converter must adhere to the requirements such as less cost, low weight, reduced switching voltage stress, and high-power density (Gulus et al., 2014).

The converters are classified into two types, namely, isolated and nonisolated based high-step-up converter. The isolated converters are magnetic coupling based converter and nonisolated converters are transformer less. This chapter is organized as follows: Section 5.2 discusses the nonisolated converter and its type. Section 5.3 deals with the isolated converters. The bidirectional converter-based isolated and nonisolated methods were illustrated in Section 5.4. The three-port DC–DC converter and its types are discussed in Section 5.5 and appropriate conclusions drawn from the study are discussed in Section 5.6.

5.2 NONISOLATED DC–DC CONVERTER

The nonisolated converters are formed based on the conventional boost, buck, and buck–boost converters, the generating voltage gain of these converters are restricted. Among nonisolated converter family, the classical boost converter generates the high-step-up voltage gain by using large duty cycle. This may cause the high switching voltage stress with reverse recovery issues and reduces the conversion efficiency (Hsieh et al., 2013 and Li and He, 2011). Many researchers concentrate on single-stage converters. To overcome above issues, the necessary requirements are high step-up, low cost, and high efficiency needed for the converters. To obtain above, the following methods can be useful:

- high-step-up converters with coupled inductor,
- high-step-up converters with switched capacitor,
- high-step-up converters with an inductor and switched capacitor,
- high-step-up converters with coupled inductor and switched capacitor, and
- high-step-up interleaved boost converters.

5.2.1 HIGH-STEP-UP DC–DC CONVERTERS WITH COUPLED INDUCTOR

The high-step-up converter efficiency is improved by using coupled inductor which generates high voltage gain, by adjusting the turn ratio of coupled inductor similar to isolated converters (Zhao and Lee, 2003). The high-step-up boost converter with coupled inductor is shown in Figure 5.1. The coupled inductor secondary winding part operates as a voltage source, which is series with main supply. However, the main switch is affected by voltage stress and high voltage spike due to the leakage energy of coupled inductor. By absorbing the leakage energy and suppressing the switch turnoff voltage by using clamp diode D_c and capacitor C_c, the voltage stress gets reduced.

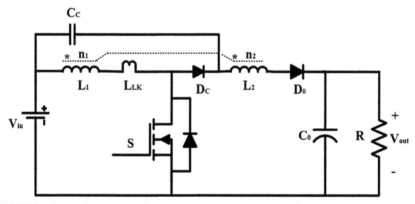

FIGURE 5.1   High-step-up DC–DC converter with a coupled inductor.

A high-step-up DC–DC converter with a coupled inductor and auxiliary circuit is presented by Wai and Lin (2005)to attain the soft-switching operation of the switch and shown in Figure 5.2a. The switch operates using the zero-voltage switching (ZVS) and zero-current switching (ZCS) turn-on conditions from the auxiliary circuit. The turn-on period of the auxiliary switches is very short in reducing additional losses. So, the converter has a complex structure and increased cost. Figure 5.2b shows a high-step-up zero-voltage transition (ZVT) of a boost converter with a coupled inductor (Wu et al., 2008). The ZVT soft switching with resonant inductor L_s is used for the main and clamp switches. To restrain the voltage spikes and to reuse the leakage energy, clamp circuit is used. The conversion efficiency of the converter is increased by utilizing the soft-switching operation.

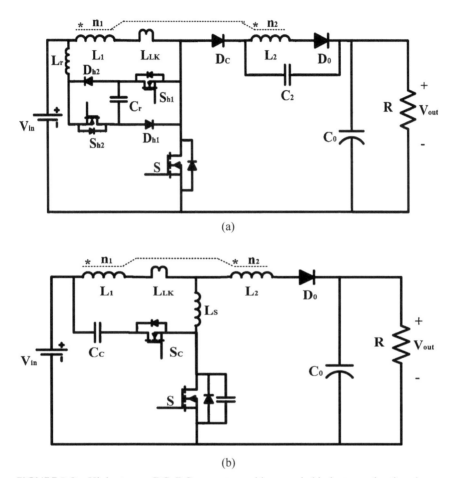

(a)

(b)

FIGURE 5.2   High-step-up DC–DC converters with a coupled inductor and active clamp.

5.2.2   HIGH-STEP-UP DC–DC CONVERTERS WITH SWITCHED CAPACITOR

Another method used to attain a high-step-up conversion is by using capacitor as a voltage source. In addition, by increasing switched capacitor in converters the high conversion ratio can be obtained. The high-step-up converter with N stage switched capacitor is shown in Figure 5.3 (Chung et al., 2003). Each switched capacitor cell is formed by a capacitor, two switches, and a diode. Each capacitor can be considered as a voltage source, which is recombined by the switches. When the switch is turned off, the diode becomes forward bias and current flows through the circuit. For increasing the high-voltage conversion in converters, N stage of switched capacitor cells are connected in series.

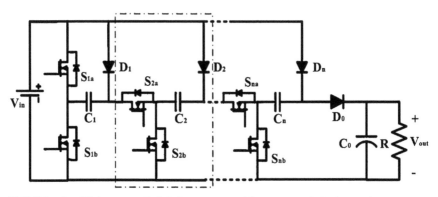

FIGURE 5.3   High-step-up DC–DC converter with N stage switched capacitors.

To limit the numbers of switches and drivers, a resonant converter-based N stage switched capacitor circuit is used as shown in Figure 5.4. The switched capacitor cell is organized by using two capacitors and diodes (Law et al., 2005). To reduce structure of the circuit, no switches were utilized in the switched capacitor cell. The switches are operated by using ZCS method of resonant tank (inductor L_r and switched capacitor) to remove the current spike issues which usually occur in classical switched capacitor converter.

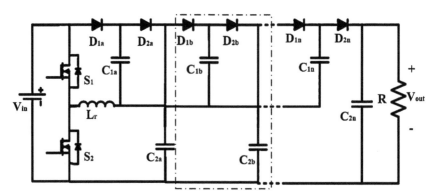

FIGURE 5.4   High-step-up DC–DC with N stage switched capacitor resonant converter.

5.2.3   HIGH-STEP-UP DC–DC CONVERTER WITH INDUCTOR AND SWITCHED CAPACITOR

The boost converter is integrated with switched capacitor to obtain a voltage gain (Ismail et al., 2008). Figure 5.5 shows a single-switch DC–DC converter with boost multiplier cell and capacitor–diode multiplier. The boost multiplier cell consists of diodes D_x and D_y, inductor L_x, and capacitor C_x which can generate the voltage conversion as a classical boost converter. The capacitor–diode multiplier consists of diodes D_1 and D_2 and the capacitors C_1 and C_2 which are used to maximize the voltage gain and to reduce the voltage stress of switch.

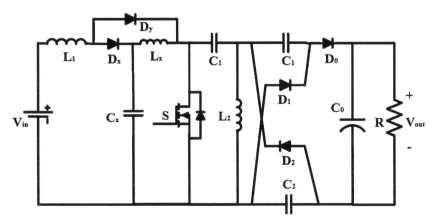

FIGURE 5.5   High-step-up with single switch converter.

The major demerits of this type high-step-up converter are: it uses hard switching which causes high switching loss and increases the number of magnetic components which limits power level. Hence, this converter type is mostly used for low power applications.

5.2.4   HIGH-STEP-UP DC–DC CONVERTER WITH SWITCHED CAPACITOR AND COUPLED INDUCTOR

The combination of coupled inductor with switched capacitor in boost converter is shown in Figure 5.6. The large voltage gain conversion can be achieved by using the converter proposed by Wai and Duan (2005). The leakage inductance of the coupled inductor causes the reverse recovery issue of the semiconductor devices. The soft-switching operation is used to reduce the switching loss. The clamped diode D_{C1} and capacitor C_{C1} are used to suppress the switch voltage and recycle the leakage energy. The resonance circuit is composed by inductor L_r and capacitor C_{C2}, which is used to transfer the capacitor C_{C1} energy to the load. The high voltage gain is obtained with reduced switch voltage stress in the converter.

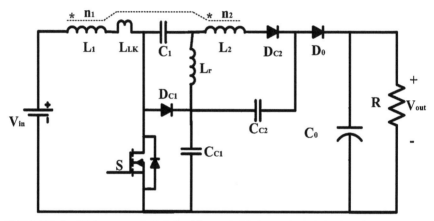

FIGURE 5.6   High-step-up DC–DC with coupled inductor-based single switch converter.

5.2.5   HIGH-STEP-UP DC–DC INTERLEAVED BOOST CONVERTERS

The interleaved structure can be useful for high-power application. The switched capacitor is integrated into the classical boost converter for generating the high voltage gain (Gules et al., 2003). The high-step-up interleaved DC–DC converter circuit is shown in Figure 5.7. This circuit is used for high voltage conversion, to decrease the ripple current, improve the transient response and reduce the component size. The high voltage gain conversion is obtained by switched capacitor cell. The demerit of this converter is its operation by hard switching condition.

The coupled inductor is formed by integrating two inductors which optimize the magnetic core and improve its magnetic utility. The interleaved boost converter with switched capacitor and coupled inductor (Giral et al., 2000) is shown in Figure 5.8. The soft switching based operation is used to reduce the magnetic components. However, the high-voltage-gain conversion is achieved by using switched capacitor cells.

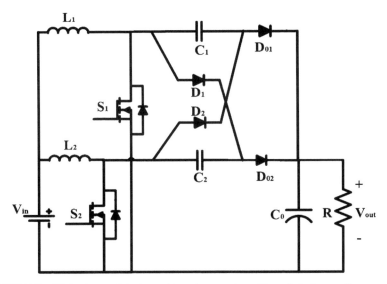

FIGURE 5.7   High-step-up interleaved boost converter with switched capacitor.

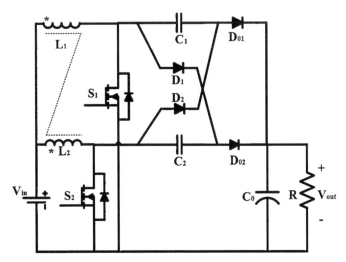

FIGURE 5.8   High-step-up interleaved DC–DC converter with coupled inductor and switched capacitor.

5.3 ISOLATED DC–DC CONVERTER

The isolated step-up converter has a transformer, and by adjusting the turn ratio of the transformer, high voltage gain is achieved. Some of those converters are flyback, push–pull, and forward-type converters. These converters have a drawback of high-voltage spike due to transformer leakage inductances. To overcome this issue diode–capacitor–resistor snubber circuits and active clamping circuits are used.

An isolated converter proposed (Spiazzi et al., 2011) by the active clamp flyback converter with a voltage multiplier to achieve the high-voltage conversion. The voltage multiplier is connected at the secondary side of the transformer as shown in Figure 5.9. This circuit overcomes the leakage inductance issue and reduces the circulating current at active clamp operation. The conduction loss of the main switch is reduced and overall efficiency is improved in the converter.

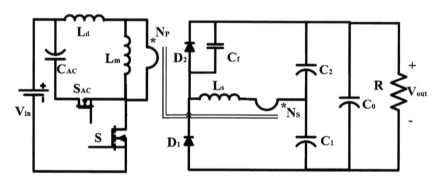

FIGURE 5.9 High-step-up series connected flyback converter with active clamp and voltage multiplier.

In a study, Lee et al. (2011) proposed an isolated DC–DC converter which is derived from series connected forward to flyback converter for high-power conversion as illustrated in Figure 5.10. This hybrid type converter shares the transformer for increasing utilization ratio. The series forward–flyback converter is connected to the secondary side of the multi-winding transformer. This converter supplies the required energy to the load through a transformer irrespective of the main switch is turned on or off. The secondary windings has low turn ratio to reduce the voltage stress of the rectifier and to improve the efficiency.

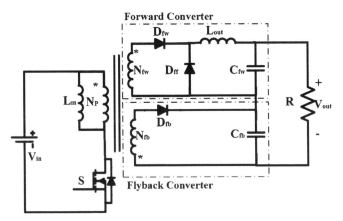

FIGURE 5.10   High-step-up series connected forward-flyback converter.

5.4   BIDIRECTIONAL DC–DC CONVERTER

The bidirectional converter operates with both the step-up and step-down principle. These types of converters are broadly used for renewable energy systems and energy storage based applications. The converters can transfer power between two DC sources in both directions. The bidirectional topologies are classified into two types, namely, nonisolated and isolated converters.

The nonisolated bidirectional converter proposed by Ardi et al. (2014) is designed using four switches and two inductors as shown in Figure 5.11. The proposed topology is simple in structure. The two switches act as power switches in both directions and other two switches act as synchronous rectifiers. The converter acts as a cascade boost/buck converter in both directions, which give lower voltage gain in step-down mode and higher gain in step-up mode. This control strategy is used in both step-up and step-down mode of operations. The coupled inductor is integrated with bidirectional DC–DC converter using four switches proposed by Duan and Lee (2012) as shown in Figure 5.12. This circuit includes soft switching and voltage clamping so as to reduce the voltage stress and conduction loss across the switches, which results in improving the conversion efficiency of the converter.

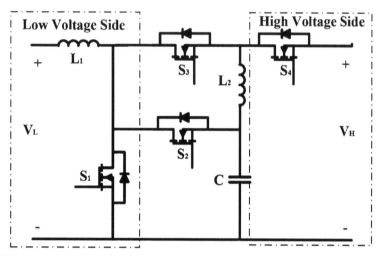

FIGURE 5.11   Nonisolated bidirectional converter.

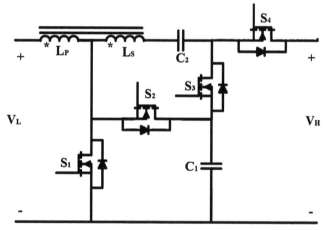

FIGURE 5.12   Nonisolated bidirectional converter with coupled inductor.

The isolated bidirectional DC–DC converters are designed by using flyback, forward, half bridge, and full-bridge-based converters. These types of converters are operated by adjusting the turn ratio of the transformer for high voltage gain in both step-up and step-down operation modes. The isolated converter proposed by Wu et al. (2010) is derived from full-bridge converter with a flyback snubber as shown in Figure 5.13. The full-bridge converter is connected to both the sides of converter with a transformer. The flyback snubber is used to recycle the absorbed energy and regulate the voltage in clamping capacitor. Thus, it reduces the current stress in full-bridge switches during full-load conditions.

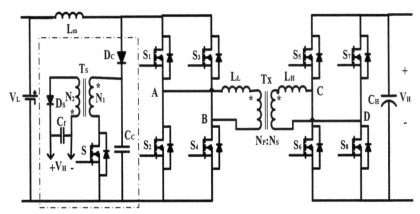

FIGURE 5.13 Isolated bidirectional converter.

5.5 THREE-PORT DC–DC CONVERTER

The main function of classical DC–DC converters is to implement the energy conversion between the input and output ports. The combination of numerous two-port DC–DC converters can be made to design multi-port converters. The goal of these converters is to make circuit simple and conversion of energy between any two of all the ports available in the converter as said in a study by Zhang et al. (2016).

The renewable energy-based power generating system utilizing a three-port DC–DC converter is shown in Figure 5.14. It consists of the DC input port which is connected to the renewable source. The energy storage system is connected in DC bidirectional portand the DC load is connected to the output port. By using power balance principle, the three-port power equation can be derived as:

$$P_{in} + P_b = P_o,$$
(5.1)

where P_{in} = input power, P_b = bidirectional power, and P_o = output power.

According to the input power condition for load demand, the three-port DC–DC converter operates in three modes of operation, as follows: when the power input is greater than the power output ($P_{in} > P_o$),the converter will work under single input dual output (SIDO) mode, the input source is renewable energy and energy storage acts as an additional load. The load demand is carried by input source and the extra power generated by source used to charge the energy storage system.

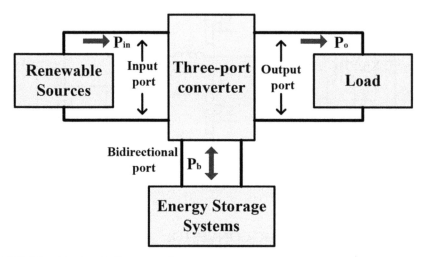

FIGURE 5.14   Block diagram of three-port converter.

When the power input is lesser than the power output ($P_{in} < P_o$), the DC–DC converter operates in the dual input single output (DISO) mode, the renewable energy and energy storage system both are connected to the input source. To support the input power, the stored energy system gets discharged to support the required load demand.

When the power input is not available or it is zero ($P_{in} = 0$), the converter will function under single input single output (SISO) mode, similar to that of the classical converter. During this mode, load demand is carried by the storage system by discharging the energy.

Owing to these modes of operation, the three-port converter provides better efficiency and larger power density. These types of converters avoid the problems occured due to the alternating nature of the renewable source and unpredictable load condition, by additional support of energy storage system. On the basis of connection of the three ports, converters are divided into three types, namely,

- nonisolated three-port converters,
- partly isolated three-port converters, and
- isolated three

5.5.1 NONISOLATED THREE-PORT CONVERTERS

Various nonisolated three-port converters have been proposed in the literature with diverse control and modulation methods. Some of the converters use a single inductor to reduce the size and further to improve the power density, whereas others use two or three inductors. Since most of these DC–DC converters are formed based on the classical boost, buck, and buck–boost converters, the voltage gain of these converters are restricted. To conquer this demerit, some three-port converters use coupled inductors to increase the ratio of voltage conversion.

A nonisolated converter with an inductor with three switches circuit for the PV application is shown in Figure 5.15. The proposed converter (Wu et al., 2013) is implemented from dual input converter by inserting new power flow with an additional control variable, which can be implemented using any one of the classical converters such as buck, boost, buck–boost, cuk, sepic, and zeta converter. The power flow of the converter is controlled by suitable control techniques.

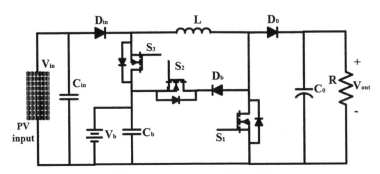

FIGURE 5.15 Nonisolated three-port converter with an inductor.

Zhu et al. (2015) proposed a new converter using three inductor and three switches as shown in Figure 5.16. This topology is formed by combining two inductor buck converter and two inductor boost converter together. To make the voltage balance of all inductors, an additional inductor has been utilized by using appropriate control strategy to operate the topology in suitable condition.

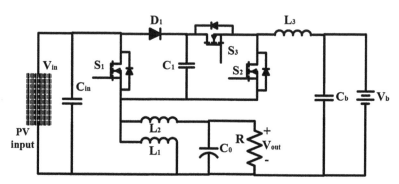

FIGURE 5.16   Nonisolated three-port converter with three inductors.

A novel three-port nonisolated converter is presented by Chien et al. (2014) as shown in Figure 5.17. By adding the coupled inductor and switched capacitor in the converter output port the high-voltage conversion with nominal duty cycle is achieved. To improve the overall efficiency of the topology, the switching voltage stresses are reduced.

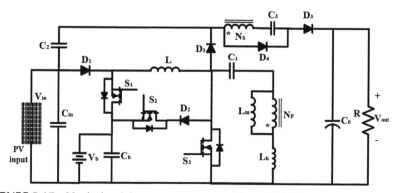

FIGURE 5.17   Nonisolated three-port converter with coupled inductor.

5.5.2 PARTLY ISOLATED THREE-PORT CONVERTERS

This type of converter usually has two ports connected directly, and then third port connected as galvanic isolation with other ports. These types of converters are generally connected by the input port (renewable source) and the bidirectional port (energy storage system) in direct form and the output port (load demand) is connected in isolated form. There are also other types of partly converters that have bidirectional port, and the output ports are connected directly and the input ports connected using galvanic isolation through a high-frequency transformer.

For renewable energy application, Wu et al. (2011) have proposed a DC–DC converter incorporated from the half-bridge converter as shown in Figure 5.18. The proposed topology have two dividing capacitors for half-bridge converter in primary side of transformer, in which one capacitor is used for the bidirectional port to allow bias DC current in the transformer to make the power flow between the input port and bidirectional port. The secondary side of the transformer is connected with two switches to perform synchronous regulation operation, which makes independent voltage regulation.

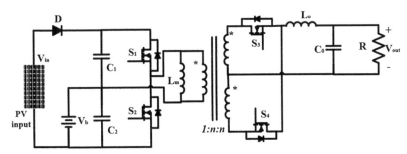

FIGURE 5.18 Partly isolated three-port converter with half bridge.

The partly isolated three-port converter is formed with a classical full-bridge converter as shown in Figure 5.19. The conventional full-bridge converter is divided into two switching legs for two individual switching cells, which are used to connect with the two different sources (Wu et al., 2012).

The proposed converter have advantages such a slow voltage stresses, all the switches in primary side of transformer use soft switching control, and appropriate controller for the power flow between two ports of the converter.

Zhu et al. (2015) proposed a new converter in which input port is isolated at the transformer primary side and the energy storage port and the load which is integrated at the transformer secondary side as shown in Figure 5.20. This converter is formed from classical half-bridge converter. The conventional half-bridge converter output is removed first and then a boost converter is inserted between the two detached outputs to provide the power flow path between the battery and the load. Finally, a boost converter is applied to the input port to minimize the ripples of the input current. The ZCS conditions are used for all the switches in the converter. Hence, switching loss is reduced and the converter overall efficiency gets increased.

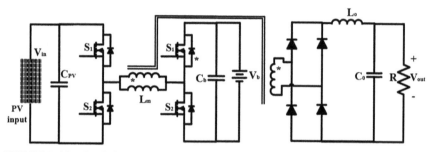

FIGURE 5.19   Partly isolated three-port converter with the full bridge.

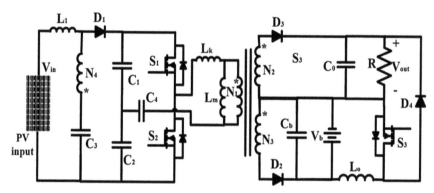

FIGURE 5.20   Partly isolated converter with half bridge and boost converter.

5.5.3   ISOLATED THREE-PORT CONVERTERS

A multiple winding high-frequency transformer which uses the power flow between any two of the three ports of DC–DC converter is called isolated converter. All three ports are connected through the galvanic isolation with their own components. Generally, these types of converters are derived from classical full-bridge converters or half-bridge converters or combination of both for conversion of energy.

The mixture of a three full-bridge circuit and three-winding transformer-based isolated DC–DC converter is presented by Duarte et al. (2007), as shown in Figure 5.21. On basis of this type of converter, many researchers have shown interest in this area, to improve the performances, control methods, and reduce its loss. In a study by Wang et al. (2012), an isolated three-port converter using half-bridge converter instead of full-bridge converter is used as shown in Figure 5.22. The input port reduces the ripple current by adding inductor and implements the soft switching to

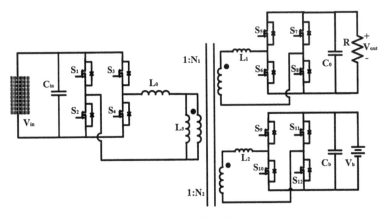

FIGURE 5.21   Isolated converter with the full bridge.

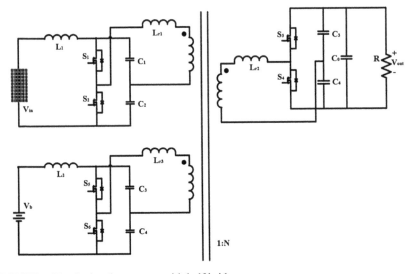

FIGURE 5.22   Isolated converter with half bridge.

diminish the switching loss, which develops the overall efficiency of the converter.

5.6   CONCLUDING REMARKS

The high-step-up DC–DC converters with various types of renewable energy sources-based applications are discussed in this chapter. The nonisolated converter topologies are discussed with different techniques used for conversion of high static gain with fast response and reduced switching voltage stress. The transformer-based high-step-up converters are explained with other methods to reduce its leakage energy and improve efficiency. The bidirectional DC–DC converter based both isolated and nonisolated converter with fewer components are described. Three-ports DC–DC converter is most suitable for hybrid energy system and its various converter topologies are illustrated. From the above converters, three-port types of high-step-up DC–DC converters are mostly suitable for renewable sources-based power production of grid application.

KEYWORDS

- renewable energy
- DC–DC converter
- bidirectional converter
- three-port converter
- isolated converter

REFERENCES

Ardi, H.; Ahrabi, R.R.; Ravadanegh, S.N. Non-isolated Bidirectional DC–DC Converter Analysis and Implementation. *IET Power Electron.* **2014,** *7*(12), 3033–3044.

Chien, L.J., et al. Novel Three-port Converter with High Voltage Gain. *IEEE Trans. Power Electron.* **2014,** *29*(9), 4693–4703.

Chung, H.S., et al. Generalized Structure of Bi-directional Switched-Capacitor dc/dc Converters. *IEEE Trans. Circuits Syst. I: Fundam. Theory Appl.* **2003,** *50*(6), 743–753.

Duan, R.-Y.; Lee, J.-D. High-efficiency Bidirectional DC–DC Converter with Coupled Inductor. *IET Power Electron.* **2012,** *5*(1), 115–123.

Duarte, J. L.; Hendrix, M.; Simoes, M.G. Three-port Bidirectional Converter for Hybrid Fuel Cell System. *IEEE Trans Power Electron.* **2007,** *22*(2), 480–487.

Giral, R., et al. Sliding-mode Control of Interleaved Boost Converters. *IEEE Trans. Circuits Syst. I: Fundam. Theory Appl.* **2000,** *47*(9), 1330–1339.

Gules, R.; Pfitscher, L. L.; Franco, L. C. In An Interleaved Boost DC–DC Converter with Large Conversion Ratio, *IEEE ISIE Proceeding,* 2003; 411–416.

Hsieh, Y.-P., et al. Novel High Step-up DC–DC Converter for Distributed Generation System. *IEEE Trans Ind. Electron.* **2013,** *60*(4), 1473–1482.

Ismail, E. H.M., et al. A Family of Single-switch PWM Converters with High Step-up Conversion Ratio. *IEEE Trans. Circuits Syst. I: Reg. Papers* **2008,** *55*(4), 1159–1171.

Ito, Y.; Zhongqing, Y.; Akagi, H. In *DC Microgrid Based Distribution Power Generation System,* the 4th International Power Electronics and Motion Control Conference, 2004; 3, 1740–1745.

Law, K.K., et al. Design and Analysis of Switched-capacitor-based Step-up Resonant Converters. *IEEE Trans. Circuits Syst. I: Fundam. Theory Appl.* **2005,** *52*(5), 1998–2016.

Lee, J.-H.; Park, J.-H.; Jeon, J.H. Series-connected Forward-flyback Converter for High Step-up Power Conversion. *IEEE Trans. Power Electron.* **2011,** *26*(12), 3629–3641.

Li, W.; He, X. Review of Nonisolated High-step-up DC/DC Converters in Photovoltaic Grid-connected Applications. *IEEE Trans. Ind. Electron.* **2011,** *58*(4), 1239–1250.

Saravanan, S.; Babu N.R. RBFN Based MPPT Algorithm for PV System with High Step-up Converter. *Energy Convers. Manag.* **2016,** *122,* 239–251.

Spiazzi, G.; Mattavelli, P.; Costabeber, A. High Step-up Ratio Flyback Converter with Active Clamp and Voltage Multiplier. *IEEE Trans. Power Electron.* **2011,** *26*(11), 3205–3214.

Sun, Y., et al. Distributed Cooperative Synchronization Strategy for Multi-bus Microgrids. *Electr. Power Energy Sys.* **2017,** *86,* 18–28.

Veneri, O.; Capasso, C.; Iannuzzi, D. Experimental Evaluation of DC Charging Architecture for Fully-electrified Low-power Two-wheeler. *Appl. Energy.* **2016,** *162,* 1428–1438.

Villalva, M.G.; Gazoli, J.R.; Filho, E.R. Comprehensive Approach to Modeling and Simulation of Photovoltaic Arrays. *IEEE Trans. Power Electron.* **2009,** *24*(5), 1198–1208.

Wai, R. J.; Duan, R. Y. High-efficiency DC/DC Converter with High Voltage Gain. *Proc. Inst. Elect. Eng. Elect. Power Appl.* **2005,** *152*(4), 793–802.

Wai, R. J.; Lin, C. Y. High-efficiency, High-step-up DC–DC Converter for Fuel-cell Generation System. *Proc. Inst. Elect. Eng. Elect. Power Appl.* **2005,** *152*(5), 1371–1378.

Wang, B.; Sechilariu, M.; Locment, F. Intelligent DC Microgrid with Smart Grid Communications: Control Strategy Consideration and Design. *IEEE Trans. Smart Grid* **2012,** *3*(4), 2148–2156.

Wang, L.; Wang, Z.; Li, H. Asymmetrical Duty Cycle Control and Decoupled Power Flow Design of a Three-port Bidirectional DC–DC Converter for Fuel Cell Vehicle Application. *IEEE Trans. Power Electron.* **2012,** *27*(2), 891–904.

Wu, T. F., et al. Boost Converter with Coupled Inductors and Buck–boost Type of Active Clamp. *IEEE Trans. Ind. Electron.* **2008,** *55*(1), 154–162.

Wu, T.-F., et al. Isolated Bidirectional Full-bridge DC–DC Converter with a Flyback Snubber. *IEEE Trans. Power Electron.* **2010,** *25*(7), 1915–1922.

Wu, H., et al. A Family of Three-port Half-bridge Converter for a Stand-alone Renewable Power System. *IEEE Trans. Power Electron.* **2011,** *26*(9), 2697–2706.

Wu, H., et al. Full-bridge Three-port Converters with Wide Input Voltage Range for Renewable Power Systems. *IEEE Trans. Power Electron.* **2012,** *27*(9), 3965–3974.

Wu, H., et al. Topology Derivation of Non-isolated Three-port DC–DC Converters from DIC and DOC. *IEEE Trans. Power Electron.* **2013,** *28*(7), 3297–3307.

Zhang, N.; Sutanto, D.; Muttaqi, K.M. A Review of Topologies of Three-port DC–DC Converters for Integration of Renewable Energy and Energy Storage System. *Renew. Sustain. Energy Rev.* **2016,** *56*, 388–401.

Zhao, Q.; Lee, F. C. High-efficiency, High Step-up DC–DC Converters, *IEEE Trans. Power Electron.* **2003,** *18*(1), 65–73.

Zhu, H., et al. A Non-isolated Three-port DC–DC Converter and Three-Domain Control Method for PV-battery Power Systems. *IEEE Trans. Ind. Electron.* **2015,** *62*(8), 4937–4947.

Zhu, H., et al. PV Isolated Three-port Converter and Energy-balancing Control Method for PV-battery Power Supply Applications. *IEEE Trans. Ind. Electron.* **2015,** *62*(6), 3595–3606.

CHAPTER 6

DESIGN AND CONTROL OF DC–AC INVERTERS

S. UMASHANKAR* and V. SRIDHAR

School of Electrical Engineering, VIT University, Vellore, Tamil Nadu, India

Corresponding author. E-mail: shankarums@gmail.com

CONTENTS

ABSTRACT

In a microgrid system, to interface the DC sources such as PV array, fuel cells, battery energy storage, etc. to an AC grid or to AC loads, it is required to have an Inverter as a power interface. When an inverter is connected to the grid, the power quality on the grid may be affected as the inverters are operated with pulse width modulation techniques. To achieve better power quality, to reduce the size and cost of output filter, control complexity, reliability and availability, the selection of inverter configuration is the main challenge. In this chapter, various inverter topologies suitable for microgrid applications are presented and comparison study on the basis of cost, control complexity, power quality, maintenance, etc. are carried out.

6.1   INTRODUCTION

Microgrid is a localized grouping of electrical sources which can be operated in synchronization with the centralized grid and also can be operated in an island-mode. In case of an emergency, it can supply the power to local loads by changing between islanded mode and grid-connected mode (Venkatraman & Khaitan, 2015). In grid-connected mode, if microgrid is capable of giving power more than local load requirement, then microgrid supplies the power to the local loads and the remaining power is transferred to the grid. In case if the power available within microgrid is less than the load requirement, it takes additional power required from the main grid. In islanded mode, the entire local load requirement needs to be supplied from various energy sources and energy storage components within the microgrid. Hence, in addition to the power generation, energy storage and power management are also the key functions to be performed in a microgrid system. Microgrids are located close to the load areas resulting in increased efficiency and reduction in transmission infrastructure. In a microgrid, DC network may consist of DC sources such as PV arrays, fuel cells, batteries, synchronous generator-based wind generator, etc. These sources are connected to DC bus through suitable converters. AC side of the microgrid, that is, point of common coupling (PCC) is connected to the sources such as diesel generator, induction generator-based wind turbines, etc. (Justo et al., 2013). Typically, a microgrid comprises low-voltage distribution systems. PCC is connected to the grid through suitable isolation transformer and breaker. The transformer is required to match the microgrid and main grid voltages and to provide the isolation.

An inverter is required to transfer power from DC side to the AC side. A grid-connected inverter system consists of power electronic converter, DC side filter, AC side filter, and AC and DC breakers connected across DC network and AC network as shown in Figure 6.1. An L-C-L filter is used at the AC terminals of the inverter to achieve sinusoidal output. A digital controller takes care of power control through the inverter. Generally for low-power and low-voltage applications, a two-level and three-phase inverter is used. Figure 6.2 shows one phase leg of two-level, three-phase inverter. This configuration consists of six insulated gate bipolar transistors (IGBT) switches with six antiparallel diodes and each single-phase inverter output is shifted by 120°. The output voltage levels are either 0 or +Vdc/−Vdc. Two modulation methods commonly used for this configuration are sinusoidal pulse width modulation and space vector pulse width modulation.

The main advantages of this configuration are: control is simple and cost of the system is less. Total harmonic distortion (THD) of the two-level inverter output without filter is around 70% (at switching frequency = 2.5 kHz, MI = 1) as shown in Figure 6.2b, hence the size of the filter required is more to achieve THD as per standards. The dv/dt at the inverter output is also high in a two-level inverter, which may cause stresses in the devices. Due to the above reasons, this configuration is not suitable for high-power applications. To improve the efficiency, performance, size, and cost of the system, it is preferred to operate the system with higher voltages. But to operate the system at higher voltages, two-level inverter needs multiple devices in series to achieve higher voltage blocking ability. To avoid the series connection of power semiconductor devices, to reduce dv/dt at the inverter output and the output filter size, multilevel inverter topologies have emerged as a potential alternative.

Figure 6.3 shows the classification of the inverters based on the source requirement for the inverter. Cascaded inverters need isolated DC sources for each phase, whereas neutral point clamped (NPC) inverters, flying capacitor, and T-type configurations operate with a common DC source. Topologies with common DC source are suitable for parallel operation, control is easy, and the cost is also less but failure in one component may stall the complete system and these configurations are not suitable for high-voltage applications. Cascaded inverters are suitable for very

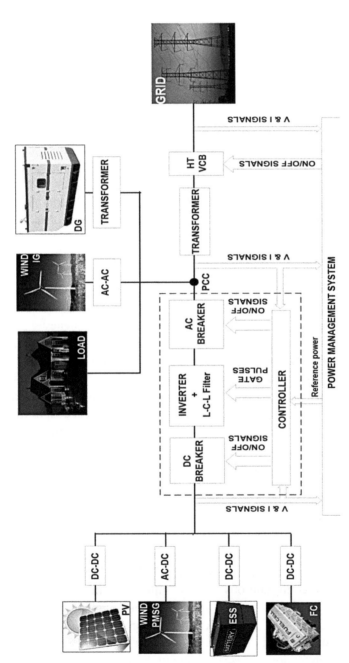

FIGURE 6.1   Block diagram of an inverter system for microgrid applications.

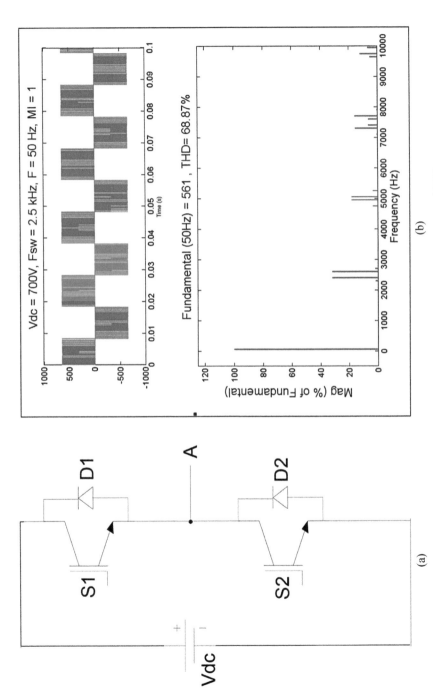

FIGURE 6.2 (a) One phase leg of two-level, three-phase inverter and (b) two-level inverter output and THD.

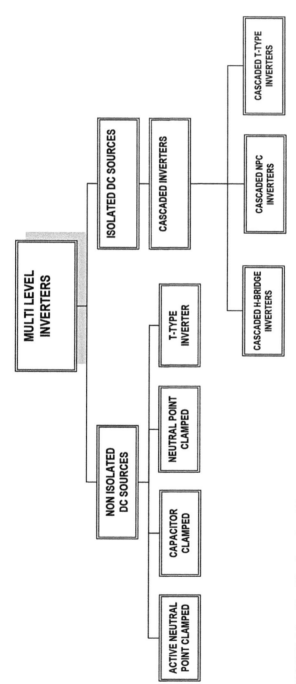

FIGURE 6.3   Classification of multilevel inverters suitable for microgrid applications based on type of DC source required.

high-power and high-voltage applications. These configurations are modular in construction, hence voltage and power rating of the system can be extended to any level by adding additional modules. In the case of failure in one module, the system can still continue to operate with a power de-rating. The cost of the system with cascaded configuration is more compared to the system with a common DC source and it needs isolated DC sources. The topologies listed in Figure 6.3 are addressed in next sections. For the comparison study, the performance of two-level inverter is taken as a reference and all the topologies are simulated for same conditions.

6.2   THREE-LEVEL NPC INVERTER

This topology is also known as diode-clamped topology and was the first step toward multilevel inverter (Rodiguez et al., 2002). One phase leg of the three-level version of this topology is shown in Figure 6.4a. This topology helped to double the inverter output voltage compared to the conventional two-level inverter without series connection of two switches. Diode-clamped topology can be extended to generate higher voltage levels. A major advantage of NPC topology is the use of a single DC link which enables using multiple inverters in parallel with a common DC link. The NPC topology shown in Figure 6.4a is a three-level topology because it can produce three output voltage levels of Vdc/2, 0, and −Vdc/2 at the terminal "A" with respect to the reference point "O." The switching strategy is such that S1 and S3 are switched on and off in a complementary fashion. Same is the case with S2 and S4.

- The Vdc/2 level is achieved when S1 and S2 are turned on. If the current 'I' is positive then both S1 and S2 conduct and connect the DC bus positive to "A." If the current is negative then both D1 and D2 conduct and connect the DC bus positive to "A." Hence, the combined voltage across S1 and S2 is Vdc. The voltage across S2 cannot exceed Vdc/2 because of the presence of D6. But voltage across S1 can be more. Hence, equalizing resistors need to be connected across S1 and S2.
- −Vdc/2 voltage level is achieved when S3 and S4 are turned on. If the current is negative then both S3 and S4 conduct and connect the DC bus negative to "A." If the current is positive then both

D3 and D4 conduct and connect the DC bus negative to "A." Here too, equalizing resistors need to be connected across S3 and S4 to ensure equal voltage distribution between them.

• Level 0: This voltage level is achieved when S2 and S3 are turned on. If the current is positive then D5 and S2 conduct and connect the DC bus midpoint to "A." If the current is negative then D4 and S3 conduct and connect the DC bus midpoint to "A." In both cases, the voltage across S1 and S3 is clamped to Vdc/2.

From the above analysis, it is clear that for a three-level NPC topology, all the devices should be capable of blocking half the DC link voltage and should be capable of carrying the peak load current. Pulse generation for three-phase inverter is exactly similar to that of the one leg inverter. The B and C phase voltage references are 120° and 240° phase shifted from the A phase voltage reference. Only one carrier is used for all the three phases. Line voltage of the inverter and the THD is shown in Figure 6.4b. THD of three-level inverter is improved compared to that of two-level inverter, hence the size of filter required is smaller with this configuration when compared with a two-level inverter.

6.3   T-TYPE ACTIVE NEUTRAL POINT CLAMPED (ANPC) INVERTER

Diode-clamped inverter configuration uses a pair of series-connected IGBTs with antiparallel free-wheeling diodes (FWD) as main switches in each arm. The current rating of all devices used in NPC is same as that in a normal two-level power converter, and the voltage rating of all the devices used in NPC is one-half of that of a two-level inverter. Switching losses, dV/dt of each arm in NPC are greatly reduced because the devices operate at one-half of the DC link voltage. However, the conduction loss of each arm in NPC is higher than that in normal two-level power converters because the current of each arm conducts through two devices in NPC instead of a single device as in a two-level inverter.

In the case of T-type configuration shown in Figure 6.5a, the same IGBTs and FWDs can be used as main switches similar to a two-level converter. A bidirectional switch is connected between the neutral point and the output terminal of each phase for the clamping to the neutral point. The current rating of all devices used in T-type configuration is same as

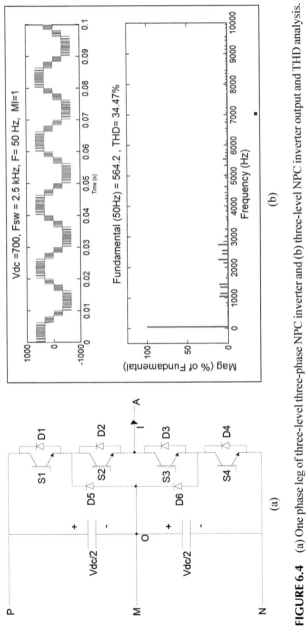

FIGURE 6.4   (a) One phase leg of three-level three-phase NPC inverter and (b) three-level NPC inverter output and THD analysis.

that in normal two-level and NPC power converters. The voltage rating of all devices used as main switches in the T-type configuration is same as that of the two-level and double that of NPC. The voltage ratings of all the devices used as clamping bidirectional switches in are one-half that of two-level and same as that of NPC. In this configuration, the switching loss and the dV/dt of each arm are greatly reduced because the devices operate under one-half of DC link voltage same as that of NPC (Schweizer & Kolar, 2013). In the case of using an antiparallel connection of a pair of series connected IGBT and diode for each bidirectional switch, the conduction loss of each bidirectional switch will be higher than that of the two-level and same as that of NPC. In the case of using a pair of anti-parallel reverse blocking IGBTs for each bidirectional switch, the conduction loss of each arm is same as that in a two-level and lower than that in the NPC because the current of each arm conducts through only a single device.

One phase leg of T-type configuration shown in Figure 6.5a is similar to the NPC inverter configuration and needs a common DC link for all the three phases and this is also a three-level topology, since it can produce three output voltage levels Vdc/2, 0, and −Vdc/2 at the terminal "A" with respect to the reference point "O."

- The Vdc/2 level is achieved when the top switch is turned on. If the current 'I' is positive then top switch conducts and connects the DC bus positive to "A." If the current is negative then D1 conducts and connects the DC bus positive to "A." Hence, the voltage across S1 and S2 is Vdc and voltage across middle IGBT is clamped to Vdc/2.

- −Vdc/2 level is achieved when bottom switch is turned on. If the current 'I' is negative then bottom switch conducts and connects the DC bus negative to "A." If the current is positive then D2 conducts and connects the DC bus negative to "A."

- Level 0: This voltage level is achieved by turning on the middle B-directional switch.

Line voltage and the THD of three-phase three-level T-type configuration is shown in Figure 6.5b. THD of three-level T-type inverter is improved compared to that of two-level inverter and remained same when compared with a three-level diode-clamped inverter.

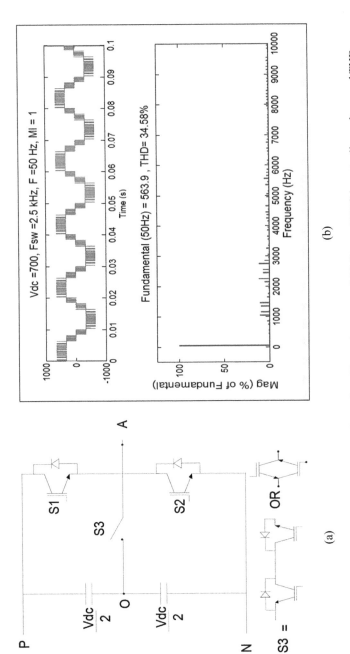

FIGURE 6.5   (a) One phase leg of three-level three-phase T-type inverter and (b) T-type ANPC inverter line voltage and THD.

6.4   FLYING CAPACITOR MULTILEVEL INVERTER

One phase leg of the five-level version of flying capacitor topology is shown in Figure 6.6a. Some of the voltage levels can be synthesized using multiple possible switching states. Hence, the switching strategy should select switching states such that the desired voltage level is produced along with balancing the flying capacitor voltages (Rodiguez et al., 2002). It is to be noted that the average power supplied by each of the flying capacitors is zero. Switches S1 and S8 are switched on and off in a complementary fashion. Similarly, switch pairs S2–S7, S3–S6 and S4–S5 also switched on and off in a complementary fashion. There are $2^4 = 16$ possible switching states and only five output voltage levels. Hence, there are multiple switching states to produce a single voltage level and for a single output voltage level. There are switching states to either charge or discharge the capacitors. With a large number of output voltage levels, the switching state multiplicity becomes so high that it becomes difficult to suitably select switching states for charge balancing of the flying capacitors. All the capacitors experience a positive and negative current of equal durations which helps in balancing the capacitors. But during current dynamics, even though durations of positive and negative currents are same, their magnitudes will be different. Hence, closed loop capacitor voltage balancing should be incorporated for proper operation of the converter. The line voltage and THD of a three-phase five-level flying capacitor inverter are shown in Figure 6.6b. THD of five-level inverter is improved compared to that of a two-level and a three-level inverter, hence the size of filter required will be much smaller when compared with a two-level inverter.

6.5   HYBRID FIVE-LEVEL INVERTER

This five-level hybrid inverter configuration is a combination of diode-clamped and capacitor-clamped inverters (Kieferndorf et al., 2010). Figure 6.7a shows one phase leg of a three-phase, five-level hybrid inverter. There are three complementary switch pairs in each of the inverter legs. The switch pairs in the leg are S1, S6; S2, S5; and S3, S4. Therefore, only three independent gate signals are required for each inverter phase. This inverter can produce an inverter phase voltage with five voltage levels across A and O.

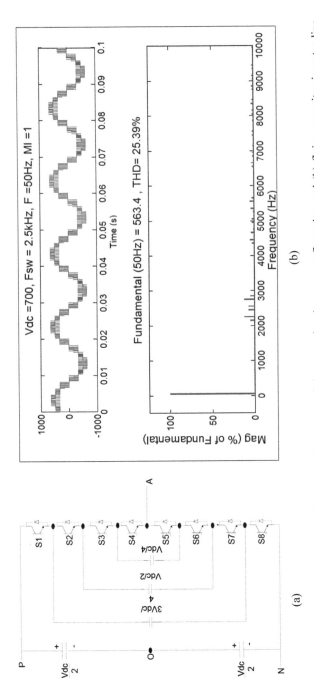

FIGURE 6.6 (a) One phase leg of five-level, three-phase flying capacitor inverter configuration and (b) flying capacitor inverter line voltage and THD.

- The Vdc/2 level is achieved when switches S1, S2, and S3 are turned on. If the current 'I' is positive then top switches conduct and connect the DC bus positive to "A." If the current is negative then D1, D2, and D3 conduct and connect the DC bus positive to "A."
- Vdc/4 level is achieved when switches S5, S6 are turned on and S4 is turned off. Vdc/4 voltage drops across the capacitor, hence the output voltage is Vdc/4.
- Level 0: This voltage level can be achieved by turning on S2, S3 or S4, S5.
- −Vdc/4 level is achieved when switches S1, S2 are turned on and S3 is turned off. Vdc/4 voltage drops across the capacitor, hence the output voltage is −Vdc/4.
- −Vdc/2 level is achieved when bottom switches are turned on. If the current 'I' is negative then bottom switch conducts and connects the DC bus negative to "A." If the current is positive then D4, D5, and D6 conduct and connect the DC bus negative to "A."

The DC capacitor voltage in this configuration normally varies with the inverter operating conditions. To avoid the problems caused by the DC voltage deviation, the voltages on the DC capacitor should be tightly controlled, which increases the complexity of the control scheme. The line voltage and THD of a three-phase five-level hybrid inverter are shown in Figure 6.7b. THD of this five-level hybrid inverter configuration is improved compared to that of a two-level and a three-level inverter and is same as five-level flying capacitor configuration.

6.6 FIVE-LEVEL CASCADED NPC OR T-TYPE MULTILEVEL INVERTER

For achieving higher voltages through NPC configuration or with T-type configuration, one option is to go for series connection of IGBTs with higher DC link voltage. But, the very idea of going for the multilevel converter is to avoid series connection and increase output voltage levels to reduce harmonics. Moreover, with the increase in DC voltage, dV/dt also increases. Another option is to use higher level diode-clamped inverter topology. But, asymmetric distribution of switching and conduction losses does not allow practical use of NPC topology having more than three levels. One solution which can be adopted is the use of three-level

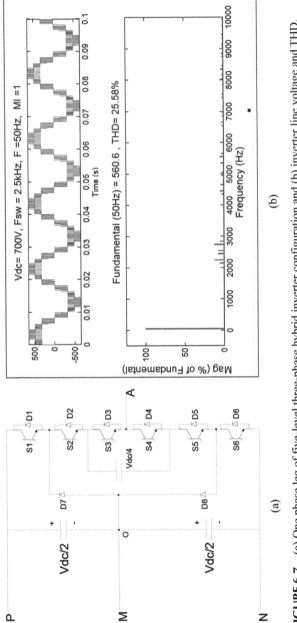

FIGURE 6.7 (a) One phase leg of five-level three-phase hybrid inverter configuration and (b) inverter line voltage and THD.

arms in a full bridge configuration to form the pole voltage. This topology can be considered as a cascaded NPC topology with three-level legs (Wanjekeche et al., 2011). Each three-level NPC full bridge can produce five levels. Five-level NPC inverter is developed by using three numbers of single-phase three-level NPC/T-type inverters, one in each phase. This inverter has some unique features that have promoted in medium-voltage applications. The inverter phase voltages, Van, Vbn, and Vcn contain five voltage levels leading to a lower dv/dt and THD. The inverter does not have any switching devices in series, which eliminates the device dynamic and static voltage sharing problems. However, the inverter requires three isolated DC supplies, which increases the complexity and cost of the DC supply system. Figure 6.8a shows a five-level cascaded NPC inverter. THD of this configuration is improved compared to that of a two-level and a three-level inverter and is same as five-level flying capacitor and five-level hybrid inverter configurations as shown in Figure 6.8b.

6.7 CASCADED H-BRIDGE MULTILEVEL INVERTER

The disadvantages of three-level NPC, that is, high dv/dt at the inverter output and the unequal power loss distribution among the devices can be mitigated by using a cascaded H-bridge (CHB) multilevel inverter topology. The concept of this inverter is based on connecting H-bridge inverters in series to get a sinusoidal voltage output. The output voltage is the sum of the voltages that is generated by each cell. The number of output voltage levels are $2n + 1$, where n is the number of cells (Wu, 2006). One phase leg of the five-level version of this topology is shown in Figure 6.9. In CHB inverter, multiple H-bridges are connected in series and every H-bridge can produce three voltage levels. Possible voltage levels across AC terminals of H-bridge-1 are Vdc, 0, and −Vdc. How the different voltage levels are synthesized in a single H-bridge, say H-bridge-1 are explained below.

- Level +Vdc is achieved when S11 and S14 are turned on. Depending on the current direction, the IGBTs or the antiparallel diodes across them conduct. Switches S12 and S13 are kept off and each of them blocks Vdc voltage.
- Level 0 is achieved when S11 and S13 or S12 and S14 are turned on together. The switches, those are off, block Vdc voltage each. When different H-bridges are series connected, a possible number of voltage levels increase.

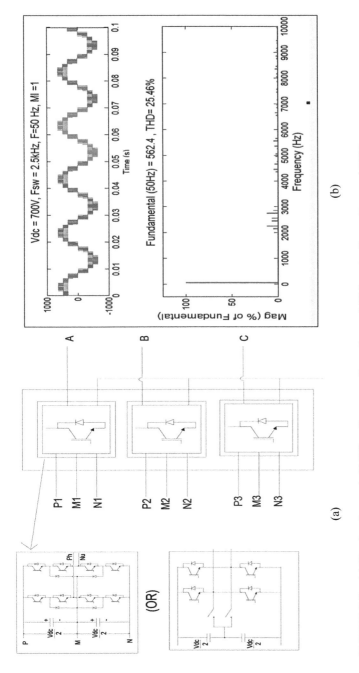

FIGURE 6.8 (a) Five-level cascaded NPC or T-type configuration and (b) inverter line voltage and THD.

- Level −Vdc is achieved when S12 and S13 are turned on. Depending on the current direction the IGBTs or the antiparallel diodes across them conduct. Switches S11 and S14 are kept off and each of them blocks Vdc voltage.

In Figure 6.9, three voltage levels Vdc, 0, and −Vdc are synthesized by each H-bridge and their combination can have five voltage levels 2Vdc, Vdc, 0, −Vdc, and −2Vdc.

- For a positive voltage reference with m < 0.5, the pole voltage swings between 0 and Vdc.
- For a positive voltage reference with m > 0.5, the pole voltage swings between Vdc and 2Vdc.
- For a negative voltage reference with m < 0.5, the pole voltage swings between 0 and −Vdc.
- For a negative voltage reference with m > 0.5, the pole voltage swings between −Vdc and −2Vdc.

The effective inverter switching frequency is four times the switching frequency of the semiconductor switches. Since number of levels achieved is five, the topology shown in Figure 6.9 is called as a five-level topology. Similarly, by cascading three H-bridges we can synthesize seven levels, by cascading four H-bridges we can synthesize nine levels and so on. Different Pulse Width Modulating (PWM) strategies are possible with CHB topology. But, phase shifted carrier-based PWM strategy is the most suitable one because it ensures equal power sharing by each module. In place of a five-level CHB topology, if a seven-level CHB shown in Figure 6.10a is selected then we get the following relations.

- For m < 1/3, the pole voltage swings between 0 and Vdc.
- For 1/3 < m < 2/3, the pole voltage swings between Vdc and 2Vdc.
- For m > 2/3, the pole voltage swings between 2Vdc and 3Vdc.

The effective inverter switching frequency will become six times the switching frequency of the semiconductor switches by using three H-bridges in cascade. The number of levels in a seven-level CHB inverter is three when the modulation index is very low, that is, less than 0.33. Number of levels is five when modulation index is between 0.33 and 0.67, and number of levels is seven when modulation index is more than 0.67.

So, it is understood that number of levels is more when the system is operated near to modulation index 1; hence, the system needs to be designed accordingly to maintain optimum modulation index for achieving more number of levels at the output voltage.

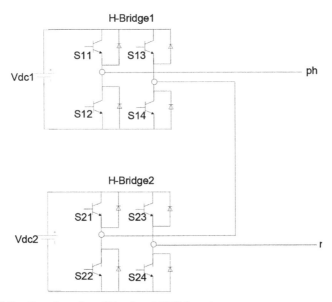

FIGURE 6.9   One phase leg of five-level CHB inverter.

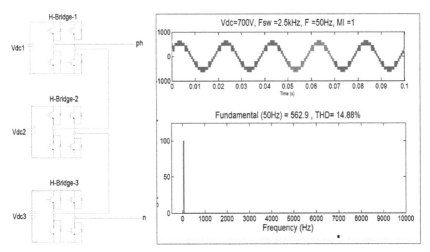

FIGURE 6.10   (a) One phase leg of seven-level CHB inverter and (b) inverter line voltage and THD.

6.8   COMPARISON OF VARIOUS INVERTER CONFIGURATIONS

The fundamental voltage of all the H-bridge cells is equal and since they are connected in series, all the H-bridges carry equal current. Hence, the power delivered by each cell is same. Because of equal DC link, equal current, equal switching frequency, and equal power factor of operation, the power loss in each cell is also same. Hence, CHB topology with phase shifted carrier is ideal for modular power converter structure. But the advantage of CHB topology is truly exploited when we use more number of voltage levels, while maintaining equal power distribution among the H-bridge cells. Figure 6.8b shows the line voltage and THD of a seven-level cascaded NPC inverter. THD of this configuration is improved compared to that of all other configurations discussed earlier.

6.9   ACTIVE AND REACTIVE POWER CONTROL

Figure 6.11 shows a detailed controller block diagram suitable for all three-phase inverters for microgrid applications. In a microgrid, power demand management system analyzes the power available with various sources and the power required by the loads. It also switches the microgrid system from grid-connected mode to islanding mode and vice versa. Power management system generates reference power to the inverter. Three-phase voltage at PCC is monitored to find out the angle ωt through phase locked loop (PLL). Angle wt obtained through PLL is used to generate Id and Iq components from three-phase inverter output currents. Reference direct axis component of AC current (Id_Ref) is calculated from the active power to be transferred and quadrature axis component of AC current (Iq_ref) is obtained from the reactive power reference. After comparing the reference Idq currents and actual Idq currents, the error signals are given to proportional integral (PI) controllers for the power control. The PI controller outputs are converted back to three-phase modulating signals and given to PWM generator to generate inverter gate pulses. PWM generator block is different for each inverter configuration discussed in earlier sections. Other than PWM generator block, the control philosophy for all the inverter configurations remains same.

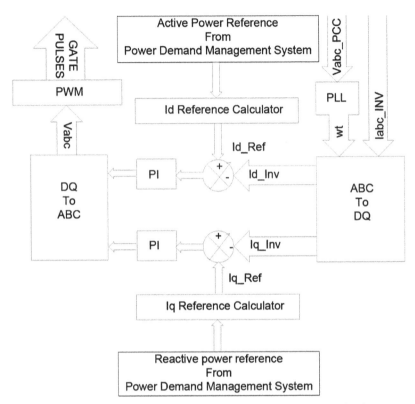

FIGURE 6.11   Control block diagram of an inverter for microgrid applications.

6.10 CONCLUDING REMARKS

In this chapter, various inverter topologies suitable for microgrid applications are discussed and comparison study is carried out. In the case of single DC source, three-level NPC inverter is more suitable than other topologies, since the output THD and filter size are less and control is simple. In the case of the availability of multiple isolated DC sources, cascaded inverters are more suitable. Based on Table 6.1, CHB topology with two-level modules shows better input and output performances than other topologies. The number of IGBT and their gate drivers is maximum for this topology. However, the cost of low-voltage IGBT and its low-voltage gate driver should be attractive over HV-IGBT and its special high-voltage gate driver.

TABLE 6.1 Comparison of Various Inverter Configurations Suitable for Microgrid Applications.

	Two-level inverter	Three-level NPC/T-type inverter	Flying capacitor inverter	Five-level hybrid inverter	Five-level cascaded NPC/T-type inverter	CHB inverter
Structure of inverter	Non-modular in structure	Non-modular in structure	Non-modular in structure	Non-modular in structure	Modular in structure	Modular in structure
DC source requirement	Single DC source	Single DC source	Single DC source	Single DC source	Three isolated DC sources are required (one per phase)	Isolated DC source for each H-bridge (nine in case of three-phase, seven-level CHB inverter)
Control complexity	Simple	Simple	Complicated since the voltage balancing of capacitors needs additional control loop	Complicated since the voltage balancing of capacitors needs additional control loop.	Not complicated	Moderately complicated due to more number of devices and communication among H-bridges
Maintenance	Simple	Simple	Due to non-modular structure, In case of failure in a power module, the complete power module needs to be changed	Due to non-modular structure, In case of failure in a power module, the complete power module needs to be changed	In case of failure, one phase can be replaced with one spare module for immediate operation	In the case of failure, one H-bridge cell needs to be changed. However, the system can still be operated with a power degradation
Advantages	*Less cost *Control is easy *Troubleshooting is easy	*Control is easy *Less filter size *Troubleshooting is easy	*Less filter size *Troubleshooting is easy	*Less filter size *Troubleshooting is easy	*Voltage balancing of capacitor is easy *Troubleshooting is easy *Suitable for high-power ratings *Due to modular structure power module design is easy	*Near to sinusoidal output. *System can be upgraded to any power and voltage levels by adding additional H-bridges *In the case of failure in a cell of one power module, the system can still be operated at lower power just by bypassing that particular cell through software

TABLE 6.1 *(Continued)*

	Two-level inverter	Three-level NPC/T-type inverter	Flying capacitor inverter	Five-level hybrid inverter	Five-level cascaded NPC/T-type inverter	CHB inverter
Disadvantages	*Not suitable for high-power, high-voltage applications *High dV/dt *High filter size	*Needs series/paralleling of devices, hence not suitable for high-power, high-voltage applications	*Voltage balancing of capacitors is difficult *In case of failure in power module, the system needs immediate shutdown	*Voltage balancing of capacitors is difficult *In case of failure in power module, the system needs immediate shutdown	*In the case of failure in power module, the system needs an immediate shutdown. However, replacement of failed module is easy	*Other than output side, the rated voltage at all the points is low *Due to modular structure, power module design is easy. *It needs so many isolated DC sources *If only one DC source is available, independent DC–DC converters along with high-frequency isolation transformers may be required for each H-bridge, which increases the size, cost, and complexity of the system

KEYWORDS

- neutral point clamped
- T-type inverters
- cascaded inverters
- microgrid
- inverter topologies

REFERENCES

Justo, J. J.; Mwasilu, F.; Lee, J.; Jung, J. W. AC-microgrids Versus DC-microgrids with Distributed Energy Resources. *Renew. Sustain. Energy Rev.* **2013**, *24*, 387–405.

Kieferndorf, F.; Basler, M.; Serpa, L. A.; Fabian, J. H.; Coccia, A.; Scheuer, G. A. In *ANPC-5L Technology Applied to Medium Voltage Variable Speed Drives Applications*, IEEE Proceedings of SPEEDAM 2010; pp 1718–1725.

Rodiguez, J.; Lai, J. S.; Peng, F. Z. Multilevel Inverter: A Survey of Topologies, Controls, and Application. *IEEE Trans. Ind. Electron.* **2002**, *49*(4), 724–738.

Schweizer, M.; Kolar, J. W. Design and Implementation of a Highly Efficient Three-level T-type Converter for Low-voltage Applications. *IEEE Trans. Power Electron.* **2013**, *28*(2), 899–907.

Venkatraman, R.; Khaitan, S. K. In *A Survey of Techniques for Designing and Managing Microgrids*, IEEE Power & Energy Society General Meeting, 2015; pp 1–5.

Wanjekeche, T.; Nicolae, D. V.; Jimoh, A. A. *Cascaded NPC/H-Bridge Inverter with Simplified Control Strategy and Superior Harmonic Suppression*; MATLAB—A Ubiquitous Tool for the Practical Engineer; InTech:Europe, 2011; pp 233–256.

Wu, B. *High-power Converters and AC Drives*; John Wiley & Sons, Inc., Hoboken: New Jersey, 2006; pp 119–142.

CHAPTER 7

HYBRID ENERGY STORAGE: INTRODUCTION AND MANAGEMENT FOR RENEWABLE MICROGRIDS

AMJED HINA FATHIMA* and PALANISAMY KALIANNAN

School of Electrical Engineering, VIT University, Vellore 632014, Tamil Nadu, India

*Corresponding author. E-mail: hina.fathima49@gmail.com

CONTENTS

ABSTRACT

Hybrid energy storages are being pursued to enable tapping of the varied energy and power characteristics offered by diverse storage systems now available, for renewable applications. The battery–supercapacitor hybrid system provides an excellent opportunity to exploit the rapid charge/discharge characteristics of supercapacitor with the advantage of extended charge retention offered by the battery storage. This chapter details the types of energy storages available and explores the possibilities of the need and benefits of employing hybrid storage systems. A brief review on architecture and applications of hybrid storage systems in renewable energy scenario is also included. A simple case study employing a battery–supercapacitor hybrid storage for smoothing of wind power generated by a 1.5 MW wind turbine is discussed and simulated results illustrated to prove the success of operating the hybrid storage system.

7.1   INTRODUCTION

Energy storage systems (ESSs) and solutions have been under renewed focus and interest due to increased penetration of renewables into the power grids. Development of microgrids and smart grids as the future electric power system solutions has necessitated efficient and better energy storage with flexible and scalable energy and power capabilities. As such, requirements cannot be met by a single storage solution economically, more and more researchers are presenting interesting hybrid solutions with combined storage options. Hybrid ESSs thus offer simple, economical, durable, and efficient solutions with many different possible combinations. This has also eliminated the premature failures, degradations, and oversized investments of storage systems in microgrids. Accessibility to quality and reliable electric power is the aim of every developing economy of the world. Microgrids and distributed generators have made this possible by integrating eco-friendly power generating solutions to light up remote areas and closed economies like universities and industries. Advancements in the technology and research have achieved immense results in terms of new power solutions at lowered costs and improved efficiencies. Stand-alone systems integrating more than one renewable and conventional power sources with or without storage solutions have been proposed in many research works. Such hybrid renewable energy systems (HRES) offer immensely optimized and feasible solutions to control and manage the microgrids. However, many issues are yet to be addressed to overcome the difficulties of integrating renewables in such microgrids. Hence, storage solutions and devices are of great importance to improve the power deliverability of a renewable power system. This chapter attempts to familiarize the reader with an extensive understanding of the various storage solutions which will form the basis for introduction of hybrid storage systems. Following this, the readers are presented with the need, the concept, and an up-to-date survey on hybrid storage solutions currently explored with an intention for use in renewable power grids. A simple case study is also included to demonstrate the integration and energy management of a battery–supercapacitor hybrid storage system for a large wind turbine.

Storing excess energy for later use is neither a new concept nor is its applicability across various genres of electric utilization a novel idea. Physical, electrochemical, and electromagnetic ESSs have been existing over generations. Increased concern for environment forced human mankind to opt for cleaner energy production and conservation, thereby promoting the focus on energy storages. Much advancement has been recently achieved for improving the efficiencies and reducing cost of storage systems especially for renewable microgrids. Battery technology, especially, has seen immense interest with introduction of flow batteries and advanced gel-based lead–acid batteries being considered very much able for such applications. The lithium batteries which have created a revolution in portable electronics have also been considered with new polymer-based lithium batteries being proposed for high-power applications. Flywheels and supercapacitors have been developed and implemented around the world. Other technologies such as superconducting magnets and metal–air batteries are still under developmental stages.

Efficient energy conservation and utilization is the demand of the hour. Different ESSs have different abilities to store and discharge power and hence find varied forms of applications in different fields of life. Batteries usually suffer extensive damage and degradation if subjected to frequent irregular and improper operation. On the contrary, ultracapacitors and flywheels can handle such situations very well but face the issue in storing the charged energy for longer durations. Pumped hydro and compressed air storage systems have high energy capacities but also require huge investments and extensive planning for land acquisition and rehabilitation. To achieve optimal storage solutions, many researchers have proposed over the past years hybrid solutions combining two or more storage systems. The most significant motive for promoting such systems is to optimize the size and costs of the involved storage devices and improve their efficiency and lifetime. This chapter explores the concept of hybrid energy storage systems (HESSs), their need, composition, architectures, and management/control studies. The chapter is constructed as follows. Section 7.2 details the available energy storage systems of importance in renewable microgrids to present an extensive understanding on each of its characteristics and limitations. Section 7.3 describes the need and the concept for HESS. Section 7.4 presents the architectures usually followed while implementing HESS for renewable microgrids and includes a brief review on studies of HESS integrated with renewable systems. Section 7.5 is a simple case study where a battery, supercapacitor HESS is proposed for a wind turbine to enable power smoothing and better reliability. The case study implores the energy management between the HESS to enable a smooth uninterrupted power delivery from the wind turbine. Section 7.6 concludes the study with a brief description of future scope of research regarding HESS in renewables.

7.2   ESSs AND CHARACTERISTICS

A review of the various storage systems and their applications has been presented in many studies including Díaz-González et al. (2012), Mahlia et al. (2014), Luo et al. (2015), etc. Some key features of energy storage technologies available so far for microgrid applications are summarized as follows.

7.2.1   MECHANICAL STORAGE

It involves storage of energy in potential/kinetic energy. These include pumped hydro storage systems, compressed air storage, and flywheels. A pumped hydro storage utilizes the differential in height between water reservoirs at different altitudes to generate electric power. It is the most widely implemented storage technology with 99% of the world's grid energy storage being hydro. Water is pumped from the lower reservoir to the higher one when demand is low and is then released from the higher reservoir through turbines generating electric power from the kinetic energy of the flowing water (Barbour et al., 2016). Energy is also stored in the form of compressed air in underground caverns. The stored air is drawn out and passed through gas turbine power plants to produce current. The heating of air with natural gas inside the power plant can be either diabatic (heating with external source) or adiabatic (heating with the energy released during compression) (Venkataramani et al., 2016). Flywheels store the mechanical inertia of a rotating flywheel to store energy. They are extremely adapted for medium- to small-scale energy storage applications. Extensive research on flywheels has helped in improving the technology and the design in the recent past. Flywheels store kinetic energy by using the electrical energy to spin a flywheel (usually by means of a reversible motor/generator). In order to retrieve the stored energy, the process is reversed with the motor that accelerated the flywheel acting as a brake, extracting energy from the rotating flywheel.

7.2.2 ELECTRICAL STORAGE

It stores energy in either elctrostatic charges or electromagnetic fields. They include superconducting magnet energy storage (SMES) and Electric Double Layer Capacitors (EDLC) or Super Capacitor Energy Storage (SCES) also known as ultracapacitors. SMES systems store energy by generating strong magnetic fields within a superconducting coil. Energy available in the storage system is independent of the discharge rating and calculated as LI2. Ultracapacitors possess the ability of charge or discharge rapidly and hence find applications in high energy density applications such as electric vehicles.

7.2.3 ELECTROCHEMICAL STORAGE

They store excess power in the form of chemical components capable of liberating electrons through chemical reactions. These include hydrogen storage and battery energy storage systems (BESS) that is, batteries. Hydrogen storage systems store electric power in the form of hydrogen gas liberated through electrolysis in a fuel cell. The only by-products generated through the process are heat and water; hence, it is deemed to be a very clean energy. The storage system has the advantage of flexibility of transportation and expandability. Batteries are extremely popular storage technologies in the market and widely commercialized and suitable for applications ranging from power applications to hybrid electric vehicles. They are mostly available as multiple cells as series and stacks to form the desired voltage and current. Each cell has an electrolyte and positive and negative electrodes. The movement of ions inside the electrolyte causes electrons to move in the external circuit, thus providing electrical energy. The energy stored in the electrolyte depends upon the mass/volume of electrodes and the power capacity for charging/discharging depends on the contact area of electrodes and electrolyte.

Older primary batteries were nonrechargeable and cannot be recharged once drained. Reactive metals such as zinc and lithium are used in primary cells as they provide high energy density. Secondary rechargeable cells can be repeatedly charged and recharged for repeated operations. BESS is a matured technology which has seen extensive development over the years (Luo et al., 2015). Starting from lead–acid batteries which have been for over 140 years to the more recent NaS and $LiFePO_4$ batteries, there are many batteries in the market today. They vary extensively in characteristics as well as performance and accordingly find use in a variety of applications. The evolution and properties of various battery systems are outlined in Tables 7.1 and 7.2, respectively.

TABLE 7.1   History of Modern Battery Development.

Year	Name of inventor	Activity
1800	Alessandro Volta (Italy)	First voltaic cell (zinc, copper disks) was developed
1802	William Cruickshank (UK)	First electric battery capable of mass production
1836	John F. Daniell (UK)	Invention of the Daniell cell
1839	William Robert Grove (UK)	Invention of the fuel cell (H_2/O_2)
1859	Gaston Planté (France)	Invention of the lead–acid battery
1899	Waldmar Jungner (Sweden)	Invention of the nickel–cadmium battery
1970	Group effort	Development of valve-regulated lead–acid battery
1991	Sony (Japan)	Commercialization of lithium-ion battery
1996	University of Texas (USA)	Identification of Li-phosphate ($LiFePO_4$)

Source: http://batteryuniversity.com/learn/article/battery_developments.

TABLE 7.2 Properties of Batteries.

Tech.	Capital cost in $/kWh	Discharge time	Specific energy (Wh/kg)	Cycling capability, @% dod	Life (yrs)	Energy η(%)	Self-discharge (%)	Maturity
PHS	500–1500	1–24 h	>400	>15,000	30–50	70–80	Very low	Mat
CAES	100–350	1–24 h	3.5–5.5	>10,000	40	40–70%*	Small	Dev.
HESS	800–1200	s–10 h	100–1000	~100	10–30	42	0	Dev.
FESS	380–2500	s–h	20–100	~100	15–20	90	100	Comm.
SMES	Very high	ms–8 s	100–1000	–	20–30	90	10–15	U.Demo
SCES	250–350	ms–1 h	0.1–5	> 5 * 10^5	s412	75–95	20–40	Developed
Battery storage systems (BESS)								
PbSO$_4$	50–150	s–h	35–50	500–2000 @70	5–15	70–95	0.1–0.3	Comm.
Na–S	200–600	s–h	100–175	2500 @100	10–20	75–89	20	Comm.
Ni–Cd	400–2400	s–h	30–80	3500 @100	10–20	70	0.2–0.3	Comm.
Li–ion	900–1300	min–h	100–200	1500–3500 @80	14–16	75–95	~1	Comm.
VRB	600	s–10h	30–50	100–13000 @75	10–20	65–85	Very low	U.Demo
ZBB	500	s–10h	60–85	2000–2500	8–10	65–85	Very low	U.Demo
PSB	300–1000	s–10h	>400		15	60–75	Very low	U.Demo

Mat.—Mature, Dev.—Developed, Comm.—Commercially available, U.Demo—Under demonstration; *Diabatic—40%, Adibatic—71% (Fathima and Palanisamy, 2014)

The battery electrodes, electrolyte and chemical reactions, and advantages and disadvantages for all types of batteries are listed in Table 7.3. A detailed profile of the major battery storage systems available for renewable system integrations is explained in Fathima and Palanisamy (2014).

TABLE 7.3   Batteries Structural Information.

Tech.	Anode	Cathode	Electrolyte	Advantages	Disadvantages
PbSO4	Pb dioxide	Sponge lead	Sulfuric acid	Easy to install, high η, low cost, low self-discharge	Short lifetime, periodic maintenance
Ni–Cd	Ni species	Cd species	Aqueous alkali solution	Faster discharge cycles, longer lifecycle, widely available.	Highly toxic and suffer from the memory effect, need continuous maintenance (Rai 2012)
NaS	Sodium	Sulfur	Ceramic beta-Al$_2$O$_3$	Long life, good energy density, high η	Thermal management, safety (Chen et al., 2009).
Li-ion	Graphite	Lithium metal oxides	Lithium salt in an organic liquid	High energy density and specific energy, fast charge and discharge capability, high η	High self-discharge. Highly inflammable and fragile
VRB	Bromine electrolyte solution	V4+/V5+ electrolyte solution	–	Scalable, high power, long duration, power rating and the energy rating are decoupled, electrolytes can be replaced easily, fast response, no self-discharge	High operating costs, lower energy density
ZBB	Bromide solution	Zn solution	–		
PSB	Sodium bromide	Sodium pölysulfide	–		

7.2.3.1   LEAD–ACID BATTERY

The charge and discharge chemical reactions occurring in the cell are as shown below:

$$Pb + SO_4^{2-} \leftrightarrow PbSO_4 + 2e^- \tag{7.1}$$

$$PbO_2 + SO_4^{2-} + 4H^+ + 2e^- \leftrightarrow PbSO_4 + 2H_2O \tag{7.2}$$

When the battery is charging, the electrodes liberate lead sulfate (PbSO4) and on discharging, lead is restored back from the lead sulfate solution to its initial state. These types of flooded batteries require continued maintenance. Valve-regulated batteries are also being developed where the electrolyte is immobilized within an absorbent material called separator (Koohi-Kamali et al., 2013) to avoid the maintenance issues. Overtime the battery electrodes face deterioration and reduce the efficiency of the battery. Battery performance also depends on temperature factor of the surroundings.

7.2.3.2 NICKEL-BASED BATTERIES

The charge and discharge chemical reactions occurring in the cell are as shown below:

$$2NiO(OH) + Cd + 2H_2O \leftrightarrow 2Ni(OH)_2 + Cd(OH)_2 \qquad (7.3)$$

In Ni–Cd cells, while discharging, $Ni(OH)_2$ is the active material of the positive electrode, and $Cd(OH)_2$ is the active material of the negative electrode. But during the charge cycle, NiOOH is the active material of the positive electrode, and metallic Cd the active material of the negative electrode (Chen et al., 2009). Other nickel-based cells with negative electrodes made of metal alloys (as in Ni–MH) and zinc hydroxides (as in Ni–Zn) are also being used to replace cadmium due to its toxicity (Chen et al., 2009; Hadjipaschalis et al., 2009; Rai 2012; Luo et al., 2015; Zimmermann et al., 2016).

7.2.3.3 SODIUM–SULFUR BATTERIES

The charge and discharge chemical reactions occurring in the cell are as shown below:

$$2Na + 4S \leftrightarrow Na_2S_4 \qquad (7.4)$$

These batteries are more suited for high-power storages due to their significant characteristics. The ceramic beta-Al_2O_3 can be used as an electrolyte and also as a separator. Usually electrolyte is solid but this causes limitations in mobility of ions which in turn reduces the battery efficiency. They are packed in a tall cylindrical structure contained by a fixed metallic cover. These batteries require high operating temperatures in the range of 350°C to maintain the liquid form of electrodes and electrolytes in molten liquid state for reactions to facilitate the occurrence of the chemical reactions. Hence, thermal management and safety are especially a concern for these batteries. High energy and power density and energy efficiencies enable it to deliver multiple times their rated power (Shakib et al., 2008; Chen et al., 2009; Díaz-González et al., 2012; Luo et al., 2015).

7.2.3.4   LITHIUM-BASED BATTERY STORAGE SYSTEMS

Priorly used in low-power portable applications such as mobiles and laptops, they have now grown to be used for higher power supply applications too. This technology is a major contender for battery systems to be integrated with hybrid vehicles in the near future. Recent research has also proved that magnesium has more structural stability compared to lithium, thereby indicating that in the future, a slow shift from lithium to magnesium may occur (Whittingham et al., 2012). The charge and discharge chemical reactions occurring in the cell are as shown below:

$$n\text{LiCoO}_2 \leftrightarrow \text{Li}_{1-n}\text{CoO}_2 + n\text{Li}^+ + n\text{e}^-$$
$$n\text{Li}^+ + n\text{e}^- + \text{C} \leftrightarrow \text{Li}_n\text{C} \tag{7.5}$$

   i) Lithium-ion: Most recent batteries have almost 100% efficiency and improved energy densities and life cycle. With a gross production of more than 2 billion cells, they are the leading cells used in mobile applications (Díaz-González et al., 2012; Luo et al., 2015).

   ii) Lithium-polymer: Compared to lithium-ion batteries, they have lower efficiency and lifetime. Also their self-discharge limits become more dependent on temperature. Hence, operating range also narrows down excluding the lower temperatures to improve safety.

7.2.3.5   FLOW BATTERY ENERGY STORAGE SYSTEMS

Flow battery systems are more recent technology originating from late 1980s, which find increasing applications in large-scale utility systems. Reactants are stored in external tanks to inflow batteries and circulated through pumps to facilitate the chemical processes. The reactions generate electrons which are then conducted through an external framework to the targeted loads. The cathode and anode are separated by using a separator which plays a major role in avoiding any form of contamination of the participating electrolytes. They are usually ion-selective and allow restricted movement of charges around the cell. These batteries experience both oxidation and reduction reactions and hence are also known as redox batteries. During the charging process, oxidation reactions occur at the anode and reduction at cathode whereas it is reversed during the discharge process. The power capacity of the system is determined by the size of the system and since the system is scalable, the energy and power capacities are also variable. The power density of the battery depends on the rate of flow of the electrolyte fluid through the cell. More inspection needs to be done to reduce shunt currents running in the bipolar system. The most commonly used commercial flow batteries are detailed below (Díaz-González et al., 2012; Luo et al., 2015).

i)   Vanadium redox flow battery (VRB): Here, sulfuric acid solutions of vanadium ions are stored in reservoirs. Analytic reservoir houses V2+ /V3+ ions whereas the catalytic reservoir houses V2+ / V3+ ions and carbon electrodes are provided inside the cell (Divya and Østergaard 2009; Koohi-Kamali et al., 2013). The charge and discharge chemical reactions occurring in the cell are as shown below:

$$V^{2+} \leftrightarrow V^{3+} + e^- \tag{7.6}$$

ii) Zinc–Bromine Flow Battery (ZBB):The reactions are as shown in Table 7.3. Most important feature of this ZBB is it has low to no self-discharge and high recyclability. The charge and discharge chemical reactions occurring in the cell are as shown below:

$$2Br^- \leftrightarrow Br_2(aq) + 2e^-$$
$$Zn^{2+} + 2e^- \leftrightarrow Zn \tag{7.7}$$

iii) Polysulfide–Bromide Flow Battery (PSB):The charge and discharge chemical reactions occurring in the cell are as shown below:

$$3NaBr + Na_2S_4 \leftrightarrow 2Na_2S_2 + NaBr_3 \qquad (7.8)$$

7.3   NEED FOR HESS

Energy crisis emerging from early 1970s with increased environmental concern were the key factors for setting the foundation for development of renewable sources for electric power generation. Many challenges are to be met for successful implementation and operation of renewable systems. Renewable power systems suffer from power quality and intermittency issues and also require large investments. Different energy sources have different energy costs associated with them such as combustion heat to power (CHP) engines cost only about 1100 $/kWh, whereas fuel cells are expensive amounting to 22,000 $/kWh. Thus, setting up of hybrid renewable systems need a deep understanding and economic analysis of various available options for the fast developing micro-and smart grids. With support from government policies and cutting edge innovations, costs of renewable power generation systems are fast decreasing, thus promoting investments in green energy. Unpredictability and noncontrol-lability of renewable power sources are proving to be a huge issue while interconnecting renewable with the power grid. The grid connectors are also to ensure that any faulty disturbance or imbalance originating in the renewable systems is not transmitted to the grid. Many nations having >60% of renewable penetration are currently facing this huge challenge. Forecasting, better and improved grid management, and stricter grid code enforcements are seen as options for managing this issue. Quality issues from voltage sag/swell to harmonics to low voltage ride through (LVRT)/high voltage ride through (HVRT) can be caused in micro-/smart grids impregnating distributed energy resources and their connected power converters which may affect the quality of power delivered to the consumer and also endanger the grid. All these factors are to be foreseen and appropriate control strategies should be implemented before estab-lishing connectivity with the grid. Distributed generators must be provided with required connectivity controls to ensure smooth transition from isolated or grid-connected mode.

There are also many economical implications that need to be addressed for implementing the storage systems. Operating an ESS in a liberalized market will enable system operators to break the monopoly of power markets. Thus, market liberalization and deregulation and its further impact on generators and consumers of energy needs to be explored. Even this is lacking without proper regulatory frameworks to ensure monetary and energy transfers taking place between the grid operators and storage systems. Thus, a fitting legal design needs to be in place for modernizing and liberalization of power systems.

As can be understood, by looking at the various characteristics of the energy storage systems available, different storage systems are equipped for different applications. Hydro and compressed air energy storages can store energy for longer period of times as compared to flywheels and SMES/SCES. On the contrary, the flywheels and SCES can operate up to many thousands of charge–discharge cycles whereas batteries face degradation when subjected to frequent deep cycles. Hence, for every application, the most important factor remains the selection of the appropriate storage system. This requires a deep understanding of the available systems, their characteristics, and market prices. Each ESS has an "energy rating" and a "power rating." Its "energy rating" will depend on the capacity of the storage system to hold the surplus power over a period of time. Whereas the "power rating" refers to the charge/discharge power capacity of the ESS at any given instant. We could say that flywheels have a high power capacity but a lower energy rating. With many applications like hybrid vehicles, it is requisite for the supporting ESS to provide a wide range of storage options like high power discharge for starting the vehicle and a satisfactory energy rating for achieving an efficient mileage (Chemali et al., 2016). Similarly, in renewable systems too, the ESS has to provide enough energy rating to hold surplus energy and also be capable for discharging high power instantaneously to eliminate intermittencies and maintain power quality. In such applications, a HESS combining two or more different storage solutions to provide for both the power and energy ratings is fast becoming a common approach. Such HESS had proved to deliver the required power and energy applications.

7.4 ARCHITECTURE AND CONTROL AND ENERGY MANAGEMENT: LITERATURE SURVEY

Earlier systems implemented batteries for many renewable sources-based microgrids to serve for storage applications. But the drawbacks faced were the shorter lifetime of batteries due to irregular and intermittent charging from renewables. Also, many batteries failed during ramping applications which make it essential to size the battery with a considerably large correction factors with high investment costs. Also, a larger sized battery also suffers from partial discharges as its maximum limits are tested under extreme conditions only which are usually rare. Recent developments have introduced many new batteries and other storage systems now capable of handling extreme storage applications with high modularity and easy operability. Further, integrating different storage systems to operate together to address diverse power and energy requirements has been proposed by many HESS studies as detailed in (Chong et al., 2016). Different architectures for HESS have been explained in many studies written on emergence on HESS (Zimmermann et al., 2016 and Chong et al., 2016). The basic methods for integrating are based on the number and the connectivity of power conditioning devices connected to charge/discharge the storage system as shown in Figure 7.1.

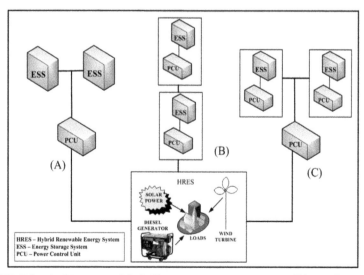

FIGURE 7.1 Architecture of hybrid energy storage systems: (A) passive parallel, (B) cascade, and (C) active parallel.

7.4.1 TYPE 1: PASSIVE PARALLEL

When two storage devices are of the equal output voltage then they can be operated in parallel without any power electronic controllers between. They may share a charge controller while connecting with the power source system. The advantage of this simple architecture is low costs, easy to integrate and lower switching losses. As there is no control on the ESS there is no opportunity to choose the operating modes of the combining ESS and hence the current and voltage sharing is uncontrolled and the ESS and power system are not isolated from each other's faults.

7.4.2 TYPE 2: CASCADED

Here the ESS are connected in series with each ESS having its own controlling power control unit which provides the protection strategy for every storage device included. This also facilitates the selection and management of the participating storage devices but in a priority based manner in the way they are connected. The limitations in the methodology exist due to increased costs and reduced scalability.

7.4.3 TYPE 3: ACTIVE PARALLEL

The HESS constructed with individual ESS units connected in parallel with each ESS equipped with its own power control unit. It provides the ultimate solution with excellent control flexibility and scalability and many studies have been proposed exploring this strategy.

There exist many studies on HESSs as listed in Prodromidis and Coutelieris (2012), Yin et al. (2014), Shao et al. (2015), Hemmati and Saboori (2016), Das et al. (2016), Faraji et al. (2017). Among all combinations of HESS battery+SC hybrid systems are most proposed as offer a perfect combination of high "power rating" and high "energy rating" for storage in renewable systems and electric vehicles (Li et al., 2010 and Wang et al., 2016). A battery-SC combined HESS was tested for a wind turbine generator using a real-time HIL simulator and the results proved the HESS to operate at better efficiencies, lower costs and improved battery life (Li et al., 2010). The SC of much lower rating combined with a larger sized battery enabled the VRB to operate with reduced power peaks and hence helped in bringing down the losses and its depth of discharge to improve its lifetime. The high power rating of the SC designed to match the power peak of the wind turbine helps in absorbing the high-frequency surges of power. A novel control strategy was presented based on operating modes for the Li-SC HESS in (Wang et al., 2016). A control study was presented for a wind-PV renewable HESS and experimental and simulated results were extracted to demonstrate the ability of the HESS to improve microgrid power quality (Tani et al., 2015).

A semi-Markov model was proposed for controlling a PV-based microgrid with a hybrid lead–acid battery-SC HESS to deliver un-intermittent PV power to the loads on a distribution system (Barnes et al., 2015). A study was presented by providing an excellent work to understand the modeling and working of a battery-SC model for a wind-PV microgrid to achieve a 65% reduction in voltage variation introduced due to uncertainties in renewable systems (Ma et al., 2015). Implementing the ESS in hybrid systems also demands intelligent and optimized control and energy management. Many studies have been proposed which implement different optimization techniques and algorithms to ensure optimal operation and cost figures for HESS. A nondominated sorting GA II (NSGA-II) algorithm was implemented to optimize the operation of a fuel-cell/battery/SC HESS (Odeim et al., 2015). Simulated annealing optimization technique was used for the HESS used in an electric vehicle based study in (Wang et al., 2010).

7.5 HYBRID ENERGY STORAGE FOR A WIND GENERATION SYSTEM—A CASE

A simple case study is simulated to study the output behavior of a large 1.5 MW Wind Turbine acting as a standalone system and feeding a constant 3-phase ac load. A block diagram of the system is shown in Figure 7.2. The wind generated power is then converted to dc via an inverter system and fed to the ac load via a dc-link and a dc-ac power converter. A hybrid storage system comprising a VRB battery and a supercapacitor (SC) is proposed to be connected with the wind turbine as shown in Figure 7.2.

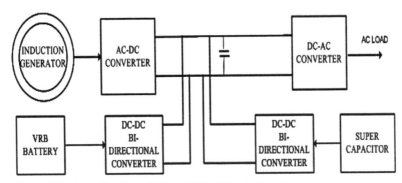

FIGURE 7.2   Block diagram of WIND-VRB-SC hybrid system.

The VRB model is developed based on the following equations (Barote et al., 2009). Stack voltage is given by,

$$V_{stack} = V_{eq} + 2\frac{RT}{F}\ln\left(\frac{SOC}{1-SOC}\right); V_{eq} = n.V_{cell} \qquad (7.9)$$

V_{eq} is calculated as the product of the number of stack cells and the individual cell voltage V_{cell}. R is the T is the temperature; F is Faradays constant and SOC is the state of charge of the VRB battery evaluated using eq. (7.10)

$$SOC(t) = SOC(t-1) \pm \frac{V \times I \times \Delta t}{E_{max}}, \qquad (7.10)$$

where V and I are the charging/discharging voltage and current to/from the battery for the duration Δt. E_{max} is the maximum battery capacity in Wh.

The wind turbine considered is a GE 1.5sle with an asynchronous machine of 575 V. The whole system is run in simulink with real time wind speed data to understand the operational capability of the modeled battery system. All necessary converters are designed for the HRES using Matlab/Simulink. The supercapacitor (SC) was selected to compensate for sudden surges and swells occurring in the system and to aid the VRB battery to ensure constant power output from the system. The SC is most suited due to its rapid response characteristics. The SC model of simulink library is used. An energy management strategy is also drafted to enable smooth power sharing among the hybrid storage systems as shown in Figure 7.3. Whenever there is a rapid change in the power difference between demand

and generation (P_{diff} exceeds battery power limits) the SC is used else the battery is operated to store surplus power or to discharge to meet energy deficits.

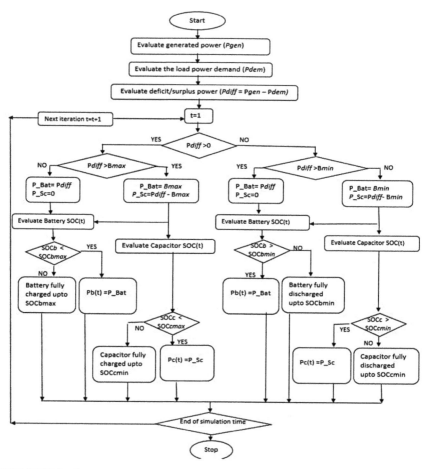

FIGURE 7.3   Energy management strategy for the hybrid VRB-SC-ESS.

The simulation is carried out for 24 h with wind speed data taken at every 30 min interval. The wind speed measured at the project site and the wind power generated is plotted as in Figure 7.4. The power difference curve shows the excess power spilled and power deficit encountered at low wind times.

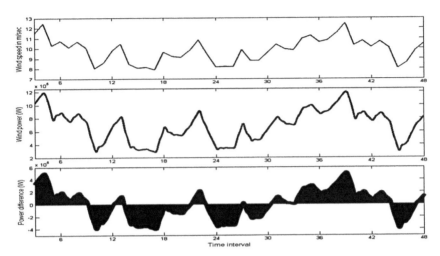

FIGURE 7.4   Power-generated curve.

The VRB is sized with a higher energy rating than the SC. Thus, the VRB battery is 1200 kWh with a power rating of 200 kW. The capacity of the SC is found as 30 kWh using the following relation, where ΔW is the energy difference which is to be delivered by the SC. Thus, the capacitance value is found to be 20 F. The SOC limits of the VRB battery are set to be 20–95% and that of the SC are 10–99%.

$$C = \frac{8}{3} \frac{\Delta W}{V_{max}^{2}} \tag{7.11}$$

The inductances and capacitances of the bi-directional buck-boost converter are calculated as below. The values thus evaluated are: VRB battery converter: L—0.015 mH; C_2, C_3—18 μF and supercapacitor converter: L—0.015 mH; C_2, C_3—28.1 μF. The simulink figures of the built models are shown in Figure 7.5.

Inductance:

$$L_2 = \frac{V_{battery} \times (V_{dclink} - V_{battery})}{I_{battery} \times fs \times V_{dclink}} \tag{7.12}$$

Buck mode capacitance:

$$C_2 = \frac{k_L \times I_{battery}}{8 \times fs \times V_{battery(ripple)}} \tag{7.13}$$

Boost mode capacitance:

$$C_3 = \frac{D_{boost} \times I_{dclink}}{fs \times V_{dclink\,(ripple)}}$$ (7.14)

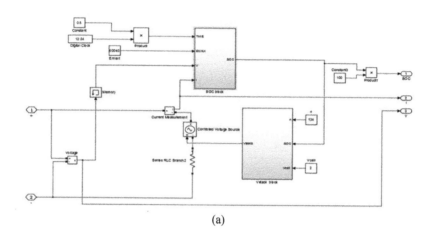

(a)

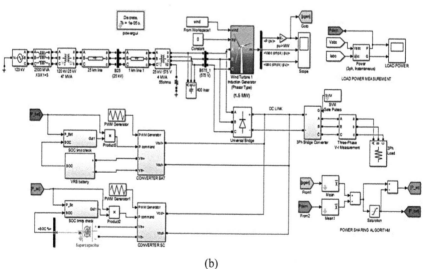

(b)

FIGURE 7.5   (a) VRB battery and (b) wind-VRB-SC HRES.

Simulation results proved successful operation of the HRES. Simulations showed the power exchange between the hybrid ESS (VRB+SC) follows the P_{diff} curve perfectly thus delivering constant power to the load. Figure 7.6 shows the power unmet curve which shows some load shedding for some 2 h when there is no power exchange from the HESS. Looking at Figures 7.7b and 7.8b, which demonstrates the SOC's of the storage systems, it is clear that the situation arose when both the VRB and the SC were completely drained of power. Figures 7.7a and 7.8a show the power exchange with the VRB and SC, respectively. It is evident that any P_{diff} upto 200 kW is handled by the VRB battery, exceeding which the SC has operated to support the VRB system.

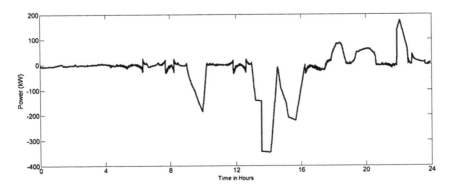

FIGURE 7.6   Power unmet curve after integrating ESS.

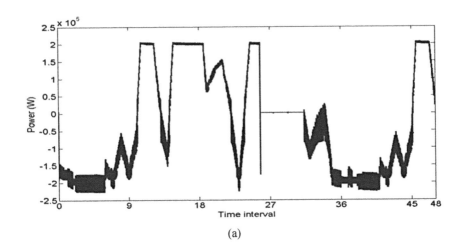

(a)

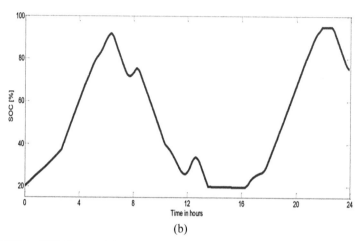

(b)

FIGURE 7.7   VRB battery (a) power exchange curve (b) SOC curve.

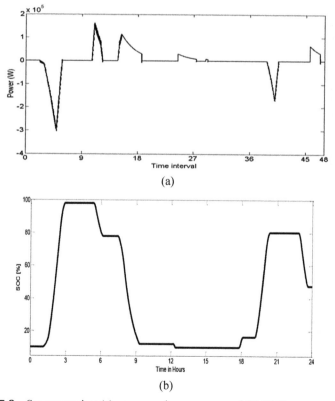

FIGURE 7.8   Supercapacitor (a) power exchange curve and (b) SOC curve.

7.6 CONCLUSION

Hybrid energy storage solutions are becoming the most sought after research and development as they offer a very flexible and practical way to plan and implement higher penetration of renewables into the power grid. Countries like India which have in the past relied heavily on hydro storage alone are also now looking for other storage options for varying applications, be it microgrids, electric vehicles or renewable integration. Further advancement and research is focused primarily on development of new materials and to overcome existing limitations. This would also help in lowering of costs and improve storage efficiencies. Optimal planning, integration and operation of storage systems are also being explored by many researchers to solve the problem of complexity introduced by the hybrid storage systems.

This study was aimed to present an overview of different storage systems, their characteristics and limitations which will form the basis for conceptualization of HESS. Accordingly, a detailed literature review on types of storage systems especially batteries was discussed. The need for opting to operate two or more storage systems was also stated in the study and an overview of the various architectures and combinations for deciding on hybrid energy storages presented. A brief case study also demonstrated the integration and management of renewable based HESSs which combines a battery and a SC to deliver uninterrupted power to the load. The battery used is a VRB which was modeled based on its SOC and an effective management strategy is evolved to ensure efficient operation of the battery to avoid premature failure. The results show satisfactory and improved performance of the hybrid system after integration of the HESS.

KEYWORDS

- energy storage
- hybrid storage
- renewable
- battery
- supercapacitor

REFERENCES

Barbour, E., et al. A Review of Pumped Hydro Energy Storage Development in Significant International Electricity Markets. *Renew. Sustain. Energy Rev.* **2016**, *61*, 421–432.

Barnes, A. K., et al. A Semi-markov Model for Control of Energy Storage in Utility Grids and Microgrids with PV Generation. *IEEE Trans. Sustain. Energy* **2015**, *6*(2), 546–556.

Barote, L., et al. In *VRB Modeling for Storage in Stand-alone Wind Energy Systems.* Proc. IEEE Bucharest Power Tech Conference, Bucharest, Romania, 2009.

Chemali, E., et al. Electrochemical and Electrostatic Energy Storage and Management Systems for Electric Drive Vehicles: State-of-the-art Review and Future Trends. *IEEE J. Emerg. Sel. Top. Power Electr.* **2016**, *4*(3), 1117–1134.

Chen, H., et al. Progress in Electrical Energy Storage System: A Critical Review. *Prog. Nat. Sci.* **2009**, *19*(3), 291–312.

Chong, L. W., et al. Hybrid Energy Storage Systems and Control Strategies for Stand-alone Renewable Energy Power Systems. *Renew. Sustain. Energy Rev.* **2016**, *66*, 174–189.

Das, H. S., et al. Proposition of a PV/Tidal Powered Micro-hydro and Diesel Hybrid System: A Southern Bangladesh Focus. *Renew. Sustain. Energy Rev.* **2016**, *53*, 1137–1148.

Díaz-González, F., et al. A Review of Energy Storage Technologies for Wind Power Applications. *Renew. Sustain. Energy Rev.* **2012**, *16*(4), 2154–2171.

Divya, K. C.; Østergaard, J. Battery Energy Storage Technology for Power Systems—An Overview. *Electr. Power Syst. Res.* **2009**, *79*(4), 511–520.

Faraji, F., et al. A Comprehensive Review of Flywheel Energy Storage System Technology. *Renew. Sustain. Energy Rev.* **2017**, *67*, 477–490.

Fathima, A. H.; Palanisamy, K. In *Battery Energy Storage Applications in Wind Integrated Systems—A Review*, Proceedings of 2014 IEEE International Conference on Smart Electric Grid (ISEG), 2014, 1–8.

Hadjipaschalis, I., et al. Overview of Current and Future Energy Storage Technologies for Electric Power Applications. *Renew. Sustain. Energy Rev.* **2009**, *13*(6), 1513–1522.

Hemmati, R.; Saboori, H. Emergence of Hybrid Energy Storage Systems in Renewable Energy and Transport Applications—A Review. *Renew. Sustain. Energy Rev.* **2016**, *65*, 11–23.

Koohi-Kamali, S., et al. Emergence of Energy Storage Technologies as the Solution for Reliable Operation of Smart Power Systems: A Review. *Renew. Sustain. Energy Rev.* **2013**, *25*, 135–165.

Li, W., et al. Real-time Simulation of a Wind Turbine Generator Coupled with a Battery Supercapacitor Energy Storage System. *IEEE Trans. Ind. Electron.* **2010**, *57*(4), 1137–1145.

Luo, X., et al. Overview of Current Development in Electrical Energy Storage Technologies and the Application Potential in Power System Operation. *Appl. Energy* **2015**, *137*, 511–536.

Ma, T., et al. Development of Hybrid Battery—Supercapacitor Energy Storage for Remote Area Renewable Energy Systems. *Appl. Energy* **2015**, *153*, 56–62.

Mahlia, T. M. I., et al. A Review of Available Methods and Development on Energy Storage; Technology Update. *Renew. Sustain. Energy Rev.* **2014**, *33*, 532–545.

Odeim, F., et al. Power Management Optimization of an Experimental Fuel Cell/Battery/Supercapacitor Hybrid System. *Energies* **2015**, *8*(7), 6302–6327.

Prodromidis, G. N.; Coutelieris, F. A. Simulations of Economical and Technical Feasibility of Battery and Flywheel Hybrid Energy Storage Systems in Autonomous Projects. *Renew. Energy* **2012**, *39*(1), 149–153.

Rai, G. D. Non-conventional Energy Sources, 4th ed.; Khanna Publishers, 2012.

Shakib, A. D., et al. In *Computation Dimensioning of a Sodium-sulfur Battery as a Backup of Large Wind Farm.* Proc. 16th PSCC, 2008, 1–6.

Shao, Q., et al. A Novel Hybrid Energy Storage Strategy Based on Flywheel and Lead-acid Battery in Wind Power Generation System. *Int. J. Control Autom.* **2015**, *8*(7), 1–12.

Sorensen, B. Renewable Energy, 3rd ed. Elsevier: Burlington, **2004**.

Tani, A., et al. Energy Management in the Decentralized Generation Systems Based on Renewable Energy—Ultracapacitors and Battery to Compensate the Wind/Load Power Fluctuations. *IEEE Trans. Ind. Appl.***2015**, *51*(2), 1817–1827.

Venkataramani, G., et al. A Review on Compressed Air Energy Storage—A Pathway for Smart Grid and Polygeneration. *Renew. Sustain. Energy Rev.* **2016**, *62*, 895–907.

Wang, B., et al. Adaptive Mode Control Strategy of a Multimode Hybrid Energy Storage System. *Energy Procedia* **2016**, *88*, 627–633.

Wang, B., et al. Adaptive Mode Switch Strategy Based on Simulated Annealing Optimization of a Multi-mode Hybrid Energy Storage System for Electric Vehicles. *Appl. Energy* **2016**. Article in Press.

Whittingham, M. S. History, Evolution, and Future Status of Energy Storage. *Proceedings of the IEEE, 100 (Special Centennial Issue)*, **2012**, 1518–1534.

Yin, J. L., et al. A Hybrid Energy Storage System Using Pump Compressed Air and Micro-hydro Turbine. *Renew. Energy* **2014**, *65*, 117–122.

Zhao, H., et al. Review of Energy Storage System for Wind Power Integration Support. *Appl. Energy* **2015**, *137*, 545–553.

Zimmermann, T., et al.; Review of System Topologies for Hybrid Electrical Energy Storage Systems. *J. Energy Storage* **2016**, *8*, 78–90.

CHAPTER 8

ENERGY FORECASTING: TECHNIQUES

N. RAMESH BABU[1]* and P. ARULMOZHIVARMAN[2]

[1]M. Kumarasamy College of Engineering, Karur, Tamil Nadu, India

[2]School of Electrical Engineering, VIT University, Vellore, Tamil Nadu, India

*Corresponding author. E-mail: nrameshme@gmail.com

CONTENTS

ABSTRACT

This chapter discusses various forecasting techniques for renewable energy such as solar and wind energy. The theoretical basics for various statistical forecasting methods are discussed in detail and validated the results with case studies. A brief discussion on numerical weather forecast models are made. To achieve the better forecast model, neural network-based forecast models have been developed and discussed in detail. Two widely used neural networks such as back propagation network and radial basis function network are used in forecast model and validated the results in the case study. The results of various models are compared and concluded the performance.

8.1 INTRODUCTION

Renewable energy technologies are the clean sources of energy which have very low impact on the environmental pollution compared with the fossil fuel technologies. In addition to that, these resources will never get exhausted such as conventional sources. In the recent years, wind energy is gaining more importance among the researchers worldwide and gained a great importance in the electric sector. Use of wind energy allows savings between 0.5 to 1 t of greenhouse gas (GHG) emissions per MWh that could be emitted to atmosphere, if thermal power plants were used instead. This contributes to a greater achievement toward the Kyoto protocol of reducing GHG emissions.

According to the grid code, IEGC (2010), the wind power generators are responsible for forecasting the daily generation with accuracy around 70% in India. But, in Europe the accuracy is around 95%. To overcome this gap, there should be an effective forecasting model to be available for the wind farm developers over short to medium term. The biggest advantage of wind power forecast is it will result in making the wind farm production similar to conventional power production by bridging the gap between the wind farm operators and the policy makers. To make the wind power more economical and feasible, it is important to maximize the efficiency of power production. Among the different aspects involved, the forecast of wind speed is one of the key factors to achieve the goal.

Malamatenios et al. (2001) state in their paper that the amount of power can be extracted from the wind is theoretically obtained by eq 8.1.

$$P = C_P \eta \frac{\rho}{2} V_1^3 A, \qquad (8.1)$$

where ρ is the air density (kg/m^3), C_P is the power coefficient, η is the mechanical or electrical efficiency, V_1 is the wind speed, and A is the rotor disk area. From eq 8.1, it is clear that the power produced by the wind mill is proportional to the cubic wind speed, which is the only varying parameter. Hence, even a small error in the forecast will increase the error to three-fold while forecast the wind power. That shows the importance of accurate wind speed forecast.

Wind speed is basically a time series data measured at regular intervals of time. On the basis of time duration of wind speed forecast, it has been classified into short-, medium-, and long-term forecasting. Short-term forecasting is an extremely important research field in the energy sector, as its time step varies from few seconds to hours, where the system operators have to handle the varying wind speed and the corresponding power generated in an optimal way. The aim of this research is to investigate the various techniques and to choose the best suitable model for forecasting the wind speed more accurately. The solar energy-based power production to electric supply is increasing tremendously. An important feature of the smart grid is its high ability to integrate renewable energy generation. The transmission system operators and utility industries are facing the fluctuating input from the photovoltaic (PV) system. This is a challenge to generate power and attain the expected load profiles. To ensure secure and economic integration of PV system into the smart grid, PV power forecasting has become important element of energy management systems. An efficient use of fluctuating energy output of PV systems requires reliable forecast information. The precise forecasting data can help to improve the quality of power delivered to the network with reduction in additional costs. The output power of PV is related to solar irradiance at the ground level, prediction of irradiance is necessary part of energy management system in the grid application (Wan et al., 2015).

The solar forecast with various conditions is considered to attain the needs of various operation and control actions, which includes grid regulation, power scheduling, and unit commitment in both distribution and transmission grids (Eftekharnejad et al., 2015). Because of the messy nature of weather systems and with uncertainties in environmental conditions such as dust, humidity, temperature, and cloud, accurate solar power forecasting can be very difficult. Various forecasting models have been implemented by many researchers in past few years for solar resources and output power of PV systems.

8.2 SOLAR FORECASTING METHODS

The forecasting methods for PV system can be classified into four models, namely,

- statistical models,
- artificial intelligence (AI) models,
- physical models, and
- hybrid models.

8.2.1 STATISTICAL MODELS

These type of models have been broadly used in time series forecasting. In general, statistical approaches are based on historical data. The forecasting aims at constructing the relationship between the variables used as inputs for the statistical model and the variable to be predicted.

8.2.1.1 PERSISTENCE

The persistence model is also known as naive predictor, broadly used for meteorology-related prediction as well as benchmark of other methods. This simple forecast method assumes that the solar irradiance/power in the future X_{t+1} will be latest measurement X_t, expressed as,

$$X_{t+1} = X_t \tag{8.2}$$

Commonly, this persistence method is the future forecast; target is average of the last T measured values described as:

$$X_{t+k} = \frac{1}{T} \sum_{i=0}^{T-1} X_{t-i}, \tag{8.3}$$

It is also called as moving average (MA). Because of its simplicity, it is mostly used as a reference model in short-term forecasting of solar energy and also in wind power (Madsen et al., 2005). The drawback of this method is the prediction accuracy which reduces drastically with forecasting horizon.

8.2.1.2 AUTOREGRESSIVE MOVING AVERAGE

Autoregressive moving average (ARMA) is one of the predominant time series prediction models due to its capability to extract useful statistical properties. It is based on two basic parts: MA and autoregressive (AR).

$$X_t = \sum_{i=1}^{p} \phi_i X_{t-i} + \sum_{i=1}^{q} \theta_i \varepsilon_{t-i}, \tag{8.4}$$

where X_t = forecasted solar irradiance/power at time t, p = order of AR model, q = order of MA error, $\varphi_i = i^{th}$ AR coefficient, $\theta_i = j^{th}$ MA coefficient, and ε = white noise, which is autonomous variable with zero mean and constant variance (Box et al., 2008).

This method is mostly used to autocorrelate time series data, and has become a practice tool for forecasting the future data of a precise time series. ARMA models are very reliable and flexible during various types of time series while using several orders, which shows better performance than the persistence method.

8.2.1.3 AUTOREGRESSIVE-INTEGRATED MOVING AVERAGE

The major demerit of ARMA method is that objective time series must be fixed, which means the time series do not change overtime in statistical properties. AR-integrated MA (ARIMA) method is used for nonstationary random processes. It is the most common class of methods for time series forecasting (Reikard 2009). The advantage of ARIMA model is having excellent capacity to capture the periodical cycle compared to other methods.

8.2.1.4 AUTOREGRESSIVE MOVING AVERAGE MODEL

In theory, both ARMA and ARIMA cannot involve the process behavior such as exogenous inputs. ARMA model (ARMAX) with exogenous inputs has proved to be an immense tool in time series forecasting. ARMAX is the extension model of ARIMA and can be more useful for real-time use of solar power forecasting because it considered external inputs such as humidity, temperature, and wind speed. It proves that the ARMAX model performances better than the other time series prediction models (Li et al., 2014).

8.2.2 AI MODELS

AI methods are used in various areas such as optimization, control, pattern recognition, forecasting, and so on. Because of its high learning and regression skills, AI models are broadly used for modeling and forecasting of solar energy.

8.2.3 PHYSICAL MODELS

Apart from statistical models and AI methods, physical models utilize PV systems to generate the solar power/irradiance forecasting. The physical methods are classified into two types, namely, sky image and numerical weather prediction (NWP) methods.

8.2.3.1 SKY IMAGE-BASED MODEL

The cloud optical depth and cover have serious influence on solar irradiance at the surface level. Considering cloud states would be helpful for solar irradiance prediction. Commonly, this method is based on analyzing the structures of cloud during a given period. The ground- and satellite-based sky image approaches have been used for forecast of local solar irradiance.

Satellite image-based methods are based on recording and detecting the structure of cloud for some period and have temporal resolution and high spatial for solar irradiance forecasting. The cloud images are used to predict GHI somewhat accurate up to 6 h ahead. The analysis data of satellite image is used to detect the motion of cloud using motion vector fields. The total sky imager (TSI) is used to detect the cloud shadow and capture the sudden changes in the solar irradiance, which used in PV power plant sites for achieving forecasting data (Chow et al., 2011).

8.2.3.2   NWP-BASED MODEL

NWP prediction method has become most precise tool for forecasting solar irradiation with look-ahead time longer than the several hours. It predicts the solar irradiance and cloud coverage percentage based on numerical dynamic modeling of the atmosphere (Lorenz et al., 2009). Commonly, NWP provides more advantages than the other forecasting models. It is used to forecast the state of atmosphere up to several days. So, it is more accurate than the satellite-based methods.

8.2.4   HYBRID MODELS

In real time, the different hybrid solar forecasting models have been proposed to improve the advantages of various types of forecasting models. Recent model combining ARMA and nonlinear ARneural network (NARNN) methods offers short-term prediction of hourly global horizontal solar irradiance and forecasting of a high resolution database using measured meteorological solar irradiance (Benmouiza and Cheknane, 2015). The combination of self-organizing maps (SOM) and hybrid exponential smoothing state space (ESSS) with artificial neural network are used in satellite-image analysis and it performs better than the conventional forecasting models (Dong et al., 2014).

8.3   WIND FORECASTING MODELS

Forecasting is defined as the projection of past into the future. It is more scientific and more objective which can be reproduced. There are different methods and models for achieving the forecast data. Mathematically, forecast can be expressed as:

$$y_t = \overline{f}\left(y_{t-1}, y_{t-2}, \ldots y_{t-d}\right), \qquad (8.5)$$

where y_t is the present data and $y_{(t-k)}$ is the past value of the series with $k = 1, 2, \ldots, d$, and $\overline{f}$ is the unknown function. One of the simplest and widely used methods for wind speed forecast is the Naïve approach.

8.3.1   NAÏVE MODEL

Naïve model forecasts for any period of time period by choosing the equivalent previous periods actual value. This model will be considered as the benchmark model by various researchers to compare and validate their model. Naïve forecast uses a single previous value of a time series data as the basis of a forecast.

$$y_t = y_{t-1},$$
(8.6)

where y_t is the actual value in period, $t-1$ and y_t are the forecast for period t.
    Other statistical models for the forecasting of time series are discussed below in detail.

8.3.2   MA MODEL

For a given series of data and the fixed subset size, the first element of MA is average of subset series. Then the subset is 'shifted forward' by excluding the first data of the series and including next data following the original subset. This creates a new subset which is averaged. The process is repeated for the whole data.

$$X_t = \mu + \varepsilon_t + \sum_{i=1}^{q} \theta_i \varepsilon_{t-i},$$
(8.7)

where θ_i is the parameter of model, μ is the mean of X_t, and ε_t and ε_{t-i} are the noise or error.
    This method is usually employed for smoothing the short-term fluctuations in the data series and it is suitable for relatively stable time series with no trend (or) seasonal pattern.

8.3.3   AR MODEL

AR model forecasts with the help of group of linear forecast formulas and attempt to forecast an output of the system based on the past outputs. Usually, this can be done by using maximum likelihood estimation.

$$\text{The forecast, } X_t = C + \sum_{i=1}^{p} \varphi_i X_{t-i} + \varepsilon_t, \tag{8.8}$$

where φ_1 and φ_p are the parameters, C is a constant, and ε_t is the white noise (random variable).

There are many ways to estimate the coefficients of the AR model such as orthogonal least squares (OLSs), methods of moments (Yale–Walker equation), or Markov Chain Monte Carlo methods as stated in Brockwell (2002). This model is prepared based on the linear assumptions and, hence, there are possibilities of leaving out the nonlinear components in the time series data and, which may result in larger error in forecasts.

8.3.4 ARMA MODEL

This is the combination of AR and MA models. The forecast can be obtained using the following equation:

$$X_t = C + \varepsilon_t + \sum_{i=1}^{p} \phi_i \varepsilon_{t-i} + \sum_{i=1}^{q} \theta_i \varepsilon_{t-i} \tag{8.9}$$

This general ARMA model is useful for low-order polynomials (of degree three or less).

In general, ARMA models can be fitted by least square regression to find the values of the parameters, which in turn minimizes the error term. And it is represented as ARMA (p and q). Appropriate values of p and q can be found by plotting partial autocorrelation functions (PACF) for estimation of p and autocorrelation functions (ACF) for estimation of q.

8.3.5 ARIMA MODEL

This is the generalization of ARMA model, by introducing the integration part in the model. The model is generally referred as ARIMA (p, d, and q), where p, d and q are nonnegative integers referring to the order of AR, integrated, and MA parts of the model, respectively. This model was invented by Box and Jenkins (1976).

The ARIMA model is expressed as:

$$\left(1 - \sum_{i=1}^{p} \varphi_i L^i\right) X_t (1-L)^d = \left(1 + \sum_{i=1}^{q} \theta_i L^i\right) \varepsilon_t, \tag{8.10}$$

where L is the log operator.

> if $d=0$, the model results in constant trend,
> if $d=1$, the model results in linear trend, and
> if $d=2$, the model results in quadratic trend.

The ease of application and reliability of the methodology has made the Box–Jenkins model the most acceptable and widely used model. Many researchers employ member of variations on ARIMA model. If the seasonal effect is expected in the forecast model, better to use seasonal ARIMA (SARIMA) and various orders on ARIMA by increasing either order of AR or MA parts.

The procedure involves three steps for determining the order of the model (p, d, and q), where p is the order of the AR component and indicates the number of AR parameters, d is the number of times the data series is differenced in order to achieve stationarity, and q is the MA order indicating the number of parameters of the MA component. The three stages involve *identification*, where values of p, d, and q are chosen; *estimation*, where coefficients of the model are obtained by employing standard statistical methods; and *diagnostic checking* of model adequacy, where the residuals of the model that were estimated at stage two of the procedure were tested for significance. A requirement of an estimation of a correct model is complete when the analysis of the residuals certifies that errors of the estimated model are independent and identically distributed, or, in other words, the error term is random and follows a white noise process.

However, while the Box–Jenkins approach allows a degree of flexibility in the choice of a model, Chatfield (2001) suggested that the flexibility also allows for a possibility to choose a misspecified model. Moreover, while original procedure required analysis of an ACF and a PACF at the identification stage, in practice it appeared to be difficult to identify the behavior of ACF and PACF of the series by comparing these plots to theoretical functions. Cho (2005) pointed out that parameters estimated by observing the ACF and PACF can be subjective, and hence lead to an unreliable and inaccurate estimation. Similarly, early studies such as Wagle (1965) considered ARMA modeling, a poor forecasting tool due to a complex estimation procedure.

However, significant improvements were made ever since in order to improve and expand the original Box–Jenkins methodology. Thus, the coefficients at the identification and estimation stages of the procedure are estimated using the Akaike's information criteria (AIC) or Bayesian information criterion (BIC), which provide more reliable statistical reference and avoid the subjectivity of the ACF and PACF interpretation. According to Brooks (2002), information criteria is a function of the residual sum of squares and accounts for the loss of degrees of freedom that occurs when extra parameters are added to the model. In the context of ARMA models specification, parameters which minimize the value of the information criteria are considered to be the correctly specified.

AIC is a measure of goodness-of-fit of an estimated model. It is expressed as:

$$AIC = -2.\ln(L) + 2(p+q+1), \qquad (8.11)$$

where L is the likelihood function.

BIC is another criterion used for model selection which attempts to correct the overfitting nature of AIC. BIC is defined as:

$$BIC = (n-p-q).\ln\left[\frac{n\hat{\sigma}^2}{n-p-q}\right] + n.(1+\ln 2\pi) + (p+q).\ln\left[\frac{\sum_{t=1}^{n} X_t^2 - n\hat{\sigma}^2}{p+q}\right], \qquad (8.12)$$

where $\hat{\sigma}^2$ is the maximum likelihood estimator of σ^2.

8.3.6   EXPONENTIAL SMOOTHING MODEL

It is the special type of MA model which includes all past observations, and use a unique set of weights from the recent observations heavily than the old observations.

If the law data is X_t, then the exponential smoothing output results as 'S_t,' which is the best estimate of next 'X' value.

$$S_1 = X_0$$

$$S_t = \alpha X_{t-1} + (1-\alpha)S_{t-1}, t = 1, \qquad (8.13)$$

where α is smoothing factor ($0<\alpha<1$).

This method is widely used, and it is suitable for relatively stable time series. Single exponential smoothing requires a minimum amount of data and computations. Hence, it is widely chosen for larger data forecast problems. Moreover, the previous error of forecast will correct the next forecast in the direction opposite to that of the error. This acts as a self-adjusting process of error correction until the minimum is reached. One major drawback in this method is the selection of the smoothing parameter. Since there is no proper procedure to choose, this trial and error approach could be applied to choose the right model for the forecast.

8.4   CASE STUDY: STATISTICAL MODEL-BASED WIND SPEED FORECAST

8.4.1   WIND DATA

To analyze the performance of the forecasting models, the wind speed dataset of samples averaged at every 15 min intervals is retrieved from a weather station located at University of Waterloo for the year 2010. The average wind speed is 3.005 m/s and the standard deviation is 1.155 m/s. The data available is measured at 3 m height, whereas standard wind speed is measured at a height of 10 m for utilizing the data in wind turbines. To obtain the wind speed at 10 m, approximately, the wind speed is multiplied by 1.5 times. This measure is based on the extrapolation of wind speed with respect to height as suggested by Malamatenios et al. (2001). To find the extrapolation factor of wind speed based on height, power law could be useful. The power law is stated as follows:

$$V = V_R \left(\frac{h}{h_R} \right)^{\alpha},$$
(8.14)

where V_R is the wind speed at reference height h_R. The exponent factor α is dependent on roughness elements of the ground, which could be obtained from area-wise sources. In general, a roughness value is chosen as (1/7).

Figure 8.1 shows the time series plot of the wind speed series of 300 samples (15 min intervals).

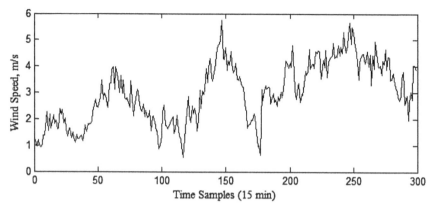

FIGURE 8.1   Wind speed data measured at 15 min intervals.

8.4.2   EVALUATION CRITERIA

The wind speed is evaluated using various criterions. The accuracy is evaluated in comparison with the actual wind speed data. The metrics used are correlation coefficient (R), coefficient of determination (R^2), mean square error (MSE), root mean square error (RMSE), mean average error (MAE), mean average percentage error (MAPE), and sum of squared error (SSE) and they are defined as follows:

The R criterion is given by,

$$R = \frac{R_{yt}}{\text{std}(y)\,\text{std}(t)}, \qquad (8.15)$$

where y and t are forecasted and actual wind speed, R_{yt} is the covariance between y and t, and std(y) and std(t) are the standard deviation of y and t, respectively. The value of R is ranging between 0 and 1 where if it is near to 1, then there is good relation between the actual and the forecast output and if it is very less or near to zero, then there exists worst forecast output in comparison with the actual data.

The R^2 criterion is given by,

$$R^2 = \left(\frac{R_{yt}}{\text{std}(y)\,\text{std}(t)} \right)^2 \qquad (8.16)$$

The MSE criterion is given by

$$\text{MSE} = \frac{1}{N}\sum_{i=1}^{N}(t_i - y_i)^2,$$ (8.17)

where N is the number of forecasted samples.

The RMSE criterion is given by

$$\text{RMSE} = \sqrt{\frac{1}{N}\sum_{i=1}^{N}(t_i - y_i)^2}$$ (8.18)

The MAE criterion is given by

$$\text{MAE} = \frac{1}{N}\sum_{i=1}^{N}|t_i - y_i|$$ (8.19)

The SSE is defined as,

$$\text{SSE} = \sum_{i=1}^{N}(t_i - y_i)^2$$ (8.20)

The MAPE is defined as,

$$\text{MAPE} = \frac{1}{N}\sum_{i=1}^{N}\left|\frac{t_i - y_i}{t_i}\right|.$$ (8.21)

8.4.3 ARMA MODEL-BASED WIND SPEED FORECAST

To find the proper forecast model of the time series using Box–Jenkins methodology, ACF and PACF is applied to the data, so as to analyze the stationarity of the time series. According to the method, for AR (p) process the sample ACF should have a decreasing appearance for AR (1) and for higher order analyze using PACF. In PACF for an AR (p) process becomes zero at lag p +1 and greater. In case of MA (q) process model, the ACF becomes zero at lag q +1, while PACF is not helpful to find q. Figures 8.2 and 8.3 show the ACF and PACF of the wind speed series.

In the plot of ACF and PACF there is no exponential decay of the graph, which indicates that there are nonstationary components available in the wind speed series. If the data are nonstationary the differencing of the time series is needed and the model order should be obtained appropriately. This inference's there is a need of analyzing the data using ARIMA model.

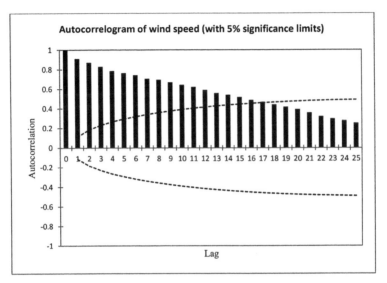

FIGURE 8.2   Autocorrelation function plot for wind speed time series.

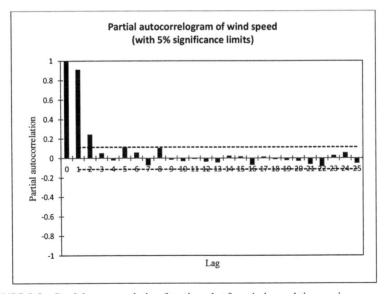

FIGURE 8.3   Partial autocorrelation function plot for wind speed time series.

To judge the model order, AIC and BIC are widely used as it they are explained in the previous section. These criterions are used to evaluate the goodness-of-fit of a parametric model.

The orders of the model areselected based on the low values of either AIC or BIC. For the AR (p) forecast model, the analysis is made to find the performance of the forecast model by increasing the order p and the results are tabulated in Table 8.1. From the table it is observed that the values of AIC and BIC areless for AR (2) model as well as the error criterions are also showing less values compared to other AR models. It can be concluded AR (2) will give good forecast among the AR models.

TABLE 8.1   AR Model Selection Parameters.

AR model	AIC	BIC	SSE	MAPE
AR (1)	376.701	384.03	61.05	14.097
AR (2)	357.83	368.745	56.747	13.679
AR (3)	358.54	373.206	56.507	13.663
AR (4)	360.47	378.807	56.494	13.684

The similar study is made for different MA models and the results are tabulated in Table 8.2. It has been observed that as the parameter is increased the values of AIC and BIC get reduced as well as the errors get reduced. This is because of the smoothing happened in the larger scale. Though the error is less for the overall series fit, there is a large error obtained in the forecast steps.

TABLE 8.2   MA Model Selection Parameters.

MA model	AIC	BIC	SSE	MAPE
MA (1)	686.809	694.14	89.445	20.634
MA (2)	582.58	593.51	31.774	12.016
MA (3)	498.37	513.03	17.796	8.571
MA (4)	468.986	487.32	7.584	5.36

The analysis is extended to the ARMA models. The AIC is less in ARMA (3, 3) model as it is shown in the Table 8.3 and major contribution is made by q parameter. The BIC analysis results for ARMA models are tabulated in Table 8.4. Table 8.5 shows the errors obtained for ARMA (p and q) models, and it is observed here that the error is also less for ARMA (3, 3). This will conclude that though error is comparatively less, the values of all the models are very near and there is no such appreciable improvement. The parameters of ARMA (3, 3) model are obtained as: C = 2.836, Φ_1 = 0.172, Φ_2 = −0.16, Φ_3 = 0.88, θ_1 = 0.483, θ_2 = 0.757, and θ_3 = −0.278. The actual wind speed and the forecasted wind speed based on ARMA (3, 3) model are shown in the Figure 8.4.

TABLE 8.3   AIC Values for ARMA Models.

AIC	MA(1)	MA(2)	MA(3)
AR(1)	356.1	358.1	359.3
AR(2)	358.1	360.1	362.1
AR(3)	360.5	361.8	355.1

TABLE 8.4   BIC Values for ARMA Models.

BIC	MA(1)	MA(2)	MA(3)
AR(1)	367.1	372.7	377.7
AR(2)	372.7	378.4	384.1
AR(3)	378.9	383.8	380.7

TABLE 8.5   Error Analysis for ARMA Models.

ARMA model	SSE	MAPE
AR (1,1)	56.416	13.682
AR (1,2)	56.413	13.684
AR (1,3)	56.26	13.709
AR (2,1)	56.411	13.686
AR (2,2)	56.416	13.682
AR (2,3)	56.409	13.691
AR (3,1)	56.503	13.629
AR (3,2)	56.355	13.705
AR (3,3)	54.464	13.404

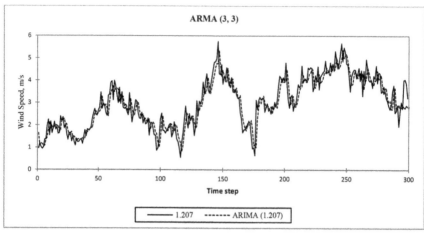

FIGURE 8.4   ARMA (3, 3) forecast output.

8.4.4   ARIMA MODEL-BASED WIND SPEED FORECAST

Since the analysis using ACF and PACF shows there is nonstationary components in the wind speed series, it has been decided to analyze using ARIMA models. Various orders of p, d, and q of ARIMA models are made and corresponding AIC and BIC values of the models were tabulated in Table 8.6. From the table it is seen that the values of AIC are low for model

TABLE 8.6   ARIMA Model Selection Parameters and Error Analysis.

ARIMA model	AIC	BIC	SSE	MAPE
1,1,1	357.23	368.22	57.06	13.416
1,1,2	358.43	373.09	56.91	13.425
1,1,3	359.48	377.79	56.714	13.399
2,1,1	357.97	372.62	56.812	13.415
2,1,2	360.17	378.48	56.85	13.421
2,1,3	360.34	382.32	56.487	13.343
3,1,1	359.93	378.24	56.8	13.415
3,1,2	361.87	383.85	54.79	13.4
3,1,3	**354.31**	**379.95**	**54.73**	**13.1**
3,2,3	372.31	397.93	57.25	13.36
3,3,3	387.76	413.35	57.94	13.08

ARIMA (3, 1, 3) which shows less error parameters as SSE value of 54.73 and MAPE of 13.1 compared to other models. In this way, the ARIMA (3, 1, 3) model is selected for the wind speed forecasting. The parameters of the ARIMA (3, 1, 3) model are obtained as: $\Phi_1 = -0.444$, $\Phi_2 = -0.626$, $\Phi_3 = 0.329$, $\theta_1 = 0.135$, $\theta_2 = 0.486$, and $\theta_3 = -0.613$. The forecasted output of this model along with the actual data is shown in the Figure 8.5.

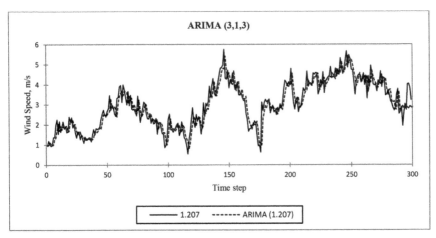

FIGURE 8.5   ARIMA (3, 1, 3) forecast output.

8.4.5 EXPONENTIAL SMOOTHING MODEL-BASED WIND SPEED FORECAST

The method which is widely used in time series forecast is exponential smoothing. As already discussed in the previous section, there is no procedure to find the best model of the simple exponential smoothing. Hence, trial and error method is applied by varying the smoothing factor within its range ($0 < \alpha < 1$). The error parameters of the various smoothing parameter models are given in Table 8.7. It is clear from the observation that the best model of exponential smoothing is obtained for the smoothing parameter (α) of 0.66. The error criterions are SSE = 57.151, MSE = 0.199, MAE = 0.345, MPE = 17.59, and the coefficient of determination, $R^2 = 0.853$. These results are better compared to other models of exponential smoothing and hence chosen as the best fit model.

TABLE 8.7   Exponential Smoothing Error Analysis.

Smoothing parameter (α)	SSE	MSE	MAE	MPE	R²
0.1	126.671	0.44	0.528	29.26	0.674
0.2	83.817	0.291	0.428	27.05	0.785
0.3	68.899	0.239	0.385	23.91	0.823
0.4	62.127	0.216	0.361	21.62	0.84
0.5	58.78	0.204	0.341	19.81	0.849
0.6	57.358	0.199	0.347	18.34	0.853
0.66	**57.151**	**0.199**	**0.345**	**17.59**	**0.853**
0.7	57.243	0.199	0.345	17.12	0.853
0.8	58.209	0.202	0.347	16.13	0.85
0.9	60.204	0.209	0.352	15.32	0.845

The forecasted output of the best exponential smoothing method along with the actual wind speed data is shown in Figure 8.6.

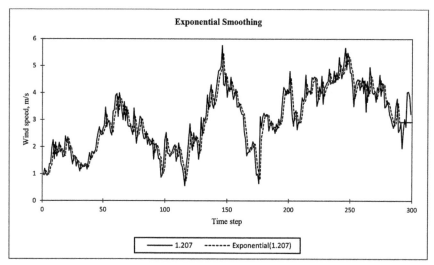

FIGURE 8.6   Exponential smoothing forecast output.

All the above discussed forecast methods are applicable to the very short period forecast only as there are assumptions to make the data as linear trend.

To overcome the limitations and to improve the performance, nonlinear models are preferred for forecasting. There are different methodologies available in the literature.

Among all neural networks are chosen as it is proven to be the best option for nonlinear mapping. The theory of the neural network and its implementation for the wind speed forecast problem has been discussed in the section 8.4 in detail and the results were compared.

8.5   STATIC NEURAL NETWORKS-BASED WIND SPEED FORECAST

8.5.1   FUNDAMENTALS OF NEURAL NETWORKS

A neural network is a computational model based on the neuron cell structure of the biological nervous system connected through simple processing units which can process the data in parallel. Given a training set of data, the neural network can learn the data with a learning algorithm.

A brief history of neural networks is as follows. The neural networks originated in 1940s. McCulloch and Pitts (1943) and Hebb (1949) have proposed the networks of simple computing devices that could model neurological activity and learning within these networks, respectively. Later, Rosenblatt (1962) has focused on computational ability in perceptrons, or single-layer feedforward networks in his study. It has been proved that perceptrons, trained with the perceptron rule on linearly separable pattern class data, could correctly separate the classes and this helps the researchers in various fields.

Later, Minsky and Papert (1988) analyzed the perceptrons and pointed out that perceptrons could not learn the class of linearly inseparable functions. It also stated that the limitations could be resolved, if networks contained more than one layer, but that no effective training algorithm for multilayer networks were available. Rumelhart et al. (1986) revived the interest in neural networks by proposing the generalized delta rule for learning by backpropagation, which is most widely used training algorithm for multilayer networks as on date.

More complex network types, different training algorithms with various application areas characterize the state-of-the-art in neural networks and its importance. Feedforward neural networks trained with backpropagation have shown revolution in this field.

Neural networks have several distinguishing features (Haykin, 2009).

- a group of processing units (neurons),
- synapses or connections between the units defined by a weight which determines the effect of the signal that process through,
- an activation function, which determines the effective input from the current node that connected to other nodes,
- an external input called bias for each unit,
- a method for information gathering usually through learning rule.

A processing unit as shown in Figure 8.7 is also called a neuron or node. The neuron receives the inputs from neighbor neurons or external sources and computes an output signal that is propagated to other units.

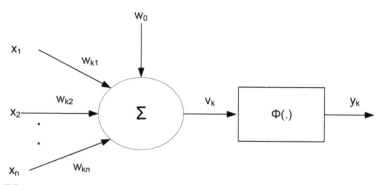

FIGURE 8.7   Processing unit.

Each unit j can have one or more inputs $x_0, x_1, x_2 \ldots x_n$, but has only one output y_k. An input to a unit is either the data from outside of the network or the output of another unit, or its own output. The neuron model includes a bias input w_0, applied externally, which effects in either increasing or decreasing the net input of the activation function based on its values.

$$v_k = \sum_{i=1}^{n} w_{ki} x_i + w_0 \qquad (8.22)$$

The contribution for positive weights, w_{ji} is considered as an excitation and an inhibition for negative weights, w_{ji}.

Most units in neural network transform their net inputs by using a scaling function called as activation function, which defines the output of the unit. Except for output units, the activation value is fed to one or more other units. There are different types of activation functions. Some of the most commonly used activation functions are:

1. threshold function,
    The output of this function is limited to one of the two values.

$$\phi(x) = \begin{cases} 1 & \text{if } (x \geq \theta) \\ 0 & \text{if } (x < \theta) \end{cases} \tag{8.23}$$

    It is also referred as Heaviside function.

2. pure linear function,

$$\varphi(x) = x \tag{8.24}$$

    It is obvious that the input units use the identity function. Sometimes a constant is multiplied by the net input to form a linear function.

3. sigmoid function

$$\phi(x) = \frac{1}{1 + e^{-x}}. \tag{8.25}$$

This function is most commonly used in neural networks, because it is easy to differentiate, and thus it can reduce the computation burden for training. It applies to applications whose desired output values are between 0 and 1. This function is commonly referred as signum function or log sigmoid function. The other form of sigmoid function which has the desired outputs varying between -1 and 1 is known as tan sigmoidal function. It has been expressed as:

$$\varphi(x) = \frac{1 - e^{-x}}{1 + e^{-x}} \tag{8.26}$$

Activation functions for the hidden units are needed to introduce nonlinearity into the networks. The reason is that a composition of linear functions is again a linear function. However, it is the nonlinearity property which makes the neural networks so powerful in its application. The sigmoid functions are the most common choices.

For the output units, activation functions should be chosen to be suited to the distribution of the target values. For continuous-valued targets with a bounded range, the sigmoid functions are again useful, provided that either the outputs or the targets to be scaled to the range of the output activation function. But, if the target values have no known bounded range, it is better to use an unbounded activation function, most often the identity function.

8.5.2 NEURAL NETWORK TOPOLOGIES

The topology of a neural network is defined by the number of layers, the number of units per layer, and the interconnection patterns between layers. Generally, the networks are classified into two categories:

1) *Feedforward* networks: In this type, the data flows from input units to output units only in the forward direction. The data processing can extend over multiple layers of units, but no feedback connections are present. Backpropagation networks (BPN) and Radial Basis Function Networks (RBFN) are the types of feedforward networks.

2) *Recurrent* networks: In this topology, the networks contain feedback connections. The connections in the network forms a cycle. In some applications in which the dynamical behavior constitutes the output of the network, the changes of the activation values of the output units are significant.

8.5.3 NETWORK LEARNING

The functionality of a neural network is determined by the combination of the topology and the weights of the connections within the network. Generally, the topology is fixed, and the weights are determined by a certain training algorithm. The process of adjusting the weights to make the network learn the relationship between the inputs and targets is called learning or training. There are different learning algorithms available to obtain an optimum set of weights that result in the appropriate solution of the problems. Broadly, the learning algorithm is divided into two main groups:

8.5.3.1 SUPERVISED LEARNING

The network is trained by providing its inputs and desired outputs (target values). The knowledge of learning between the input–output pairs is provided by an external variable, or by the system containing the network. The difference between the real outputs and the desired outputs is used by the algorithm to adapt the weights in the network.

8.5.3.2 UNSUPERVISED LEARNING

In this type of learning, there is no feedback from the environment to indicate if the outputs of the network are correct. The network itself has to reveal the features, regulations, correlations, or categories of the input data automatically.

8.5.4 FEEDFORWARD NEURAL NETWORKS

A layered feedforward network consists of a one hidden layer as shown in Figure 8.8. The network has an input layer, an output layer, and one or more hidden layers between the input and the output layer. Each unit receives its inputs directly from the previous layer and sends its output directly to units in the next layer. There are no feedback connections from any of the units to the inputs of the previous layers or to other units. Every unit only acts as an input to the immediate next layer. Obviously, this class of networks is easier to analyze theoretically than other general topologies, because their outputs can be represented with explicit functions of the inputs and the weights.

In this network there are n inputs, m hidden units, and k output units. The output of the jth hidden unit is obtained by first forming a weighted linear combination of the n input values along with the bias component, given as:

$$v_j = \sum_{i=1}^{n} w_{ji}^{(1)} x_i + w_{j0}^{(1)}, \qquad (8.27)$$

where $w_{ji}^{(1)}$ is the weight from input i to hidden unit j in the first layer and $w_{j0}^{(1)}$ is the bias for hidden unit j.

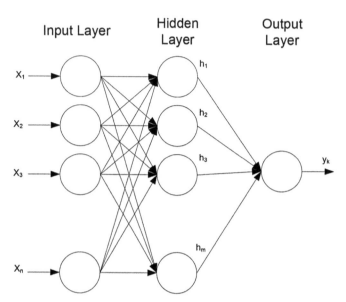

FIGURE 8.8   Feedforward neural network—architecture.

The output of the hidden unit j is obtained by passing the linear sum through an activation function $\varphi(x)$.

$$h_j = \varphi(v_j) \tag{8.28}$$

For each output unit k, first the linear combination of the output of the hidden units is formed as:

$$a_k = \sum_{j=1}^{m} w_{kj}^{(1)} h_j + w_{k0}^{(1)} \tag{8.29}$$

Then, applying the activation function $\varphi(x)$ to (4.8), we can get the kth output as:

$$y_k = \varphi(a_k) \tag{8.30}$$

The network chosen in Figure 8.8 for discussion is a network with one hidden layer. As required for the problems it can be extended to two or more hidden layers easily and compute accordingly.

Though the network exists theoretically to simulate a problem to any accuracy, there is no easy way to find it. To define an exact network architecture, such as how many hidden layers should be used and how many units should there be within a hidden layer for a certain problem is always a challenging task. Here, issues to be considered when a network is designed are discussed briefly.

It has been determined that from majority of practical problems, there is no reason to use more than one hidden layer. Practically, the problems that require two hidden layers are encountered rarely. Even for problems requiring more than one hidden layer, theoretically, most of the time, using one hidden layer is much better than using two hidden layers in practice, according to Kolmogorov (1965). The training algorithm slows down in the network when more hidden layers are used. There are many reasons to use few hidden layers as possible in practice.

1) Most training algorithms for the feedforward network are gradient based. If more hidden layers are added then the error which has to be back propagated makes the gradient very unstable. The success of any gradient-based optimization algorithm depends on the degree to which the gradient remains unchanged as the parameter varies.

2) The number of local minima increases dramatically with more hidden layers even though the algorithm is capable of finding global minima. This results in false finding or in some cases it consumes more time consuming iterations.

In general, it is strongly recommended that one hidden layer with m number of units will be the right choice for any practical feedforward network design. If that model does not perform well, then it may be worth trying a second hidden layer with fewer units compared with first hidden layer.

The next important task lies in choosing the number of units in the hidden layer. Different architectures use different number of units. Using very few units can fail to detect the signals which results in under fittingand in contrast using more number of hidden units will increase the training time and might cause overfitting.

There is no proper procedure to choose the best number of hidden units. The factors for choosing depends on the numbers of input and output units, the number of training cases, the amount of noise in the targets, the complexity of the error function, the network architecture, and the training algorithm.

There are many empirical rules for selecting the number of units in the hidden layers according to Zhang et al. (1998).

- $m \in [n, k]$ —between the input layer size and output layer size,

- $m = \dfrac{2(n+k)}{3}$ —two third of the input layer size plus the output layer size,

- $m < 2n$ —less than twice the input layer size,

- $m = \sqrt{n \cdot k}$ —squared input layer size multiplied by output layer size, and

- $m = \left(\sqrt{n+1}\right) + 10$ —squared input layer size plus 10 neurons.

Those rules stated by various researchers can only be taken as a rough reference when selecting a hidden layer size. The best approach to find the optimal number of hidden units is trial and error. It can be concluded that the above rules can be considered as an reference for trial and error approach, and fix the right choice by obtaining the performance of each case so as an optimal solution is reached for the defined problem.

8.5.5   BACKPROPAGATION ALGORITHM

Backpropagation methodology is the most widely used method for training multilayer feedforward networks. It can be applied to any feedforward network with differentiable activation functions.

For most networks, the learning process is based on a suitable error function, which is then minimized with respect to the weights and bias. The algorithm for evaluating the derivative of the error function is known as backpropagation, because it propagates the errors backward through the network and updates the weights for the next iteration.

Backpropagation algorithm performs the gradient descent method optimization of error signal and then propagates backwards iteratively so as to minimize it. The weight updating can be done in two ways: sequential update and batch update. In sequential update mode, the weights are updated for each training sample fed to the network in a single epoch, whereas the batch update mode updates the weight at the end of an epoch for all the data samples. In the real time application, the sequential mode learning is chosen.

At the nth iteration, the error signal of the output neuron, k is obtained as:

$$e_k(n) = d_k(n) - y_k^o(n), \qquad (8.31)$$

where $d_k(n)$ and $y_k^o(n)$ are the desired and the actual output of the neuron, respectively.

The instantaneous total energy of the output layer is defined as:

$$E(n) = \frac{1}{2} \sum_{k=1}^{l} e_k^2(n), \qquad (8.32)$$

where l is the number of output layer neurons.

According to gradient descent minimization algorithm, $E(n)$ should be reduced by updating the weights. The weight is updated as:

$$\Delta w_{k,j}(n) = -\eta \frac{\partial E(n)}{\partial w_{k,j}(n)}, \qquad (8.33)$$

where η is the learning rate ($0 < \eta < 1$). The next weight value (update) is done as:

$$w_{k,j}(n+1) = w_{k,j}(n) + \Delta w_{k,j}(n) \qquad (8.34)$$

The connecting weights of the output layer based on the activation function are done using:

$$\Delta w_{k,j}^o(n) = \eta \delta_k^o(n) y_j^h(n), \qquad (8.35)$$

where the local gradient $\delta_k^o(n)$ is defined by:

$$\delta_k^o(n) = e_k(n) \varphi^1(a_k(n)) \qquad (8.36)$$

The connecting weights between input and hidden layer are updated using:

$$\Delta w_{j,i}^h (n) = \eta \delta_j^h (n) x_i (n),$$ (8.37)

where the local gradient $\delta_j^h (n)$ is defined by:

$$\delta_j^h (n) = \varphi^1 \left(v_j^h (n) \right) \delta_k (n) w_{k,j}^o (n)$$ (8.38)

The gradient descent algorithm provides only the direction of the update to change, but the step size or learning rate has to be decided appropriately. Choosing the very low learning rate makes the network learning slower, while very high learning rate may lead to oscillation. One way to avoid oscillation for large learning rate η is by providing an additional factor called momentum term along with the weight change. This is expressed as:

$$\Delta w_{j,i}^{(\tau+1)} = -\eta \frac{\partial E}{\partial w_{j,i}} + \alpha \Delta w_{j,i}^{(\tau)}$$ (8.39)

Now, the weight change is a combination of a step down of the gradient term along with a fraction α of the previous weight change, where $0 \leq \alpha < 1$ and typically $0 \leq \alpha < 0.9$.

8.5.6 OTHER OPTIMIZATION ALGORITHMS

Although the gradient descent optimization method-based backpropagation learning algorithm is widely used and proven very successful in many of the applications, it does have two disadvantages such as:

1) the convergence tends to be extremely slow, and
2) convergence to the global minimum is not guaranteed.

Many researchers Bishop (1995), Fukuoka and Matsuki (1998), and Salomon and Hemmen (1996) have proposed certain improvements to the gradient descent method such as dynamically modifying learning parameters or adjusting the steepness of the sigmoid function.

In appropriate applications, other optimization methods may result in better convergence than the gradient descent and also they show a higher probability of convergence to global minima. In this section, other training algorithms which are widely used are discussed briefly.

8.5.6.1 CONJUGATE GRADIENT DESCENT ALGORITHM

Conjugategradient descentmethod proposed byFletcher (1987) is the most often recommended optimization methods to replace the gradient descent in BPN. This is based on direction set minimization method. Minimization along a direction d brings the function E to a place where its gradient is perpendicular to d. Instead of following the gradient at every step, a set of n directions is constructed which are all conjugate to each other, such that minimization along one of these directions does not spoil the minimization along one of the earlier direction. Moller (1993) and Charalambous (1992) have concluded that this method in general follows the line search to find the minimum gradient and hence the algorithm is faster in convergence.

8.5.6.2 RESILIENT BACKPROPAGATION ALGORITHM

The partial derivatives of the error function for the weight update results in small change in the magnitude in the gradient descent algorithm. This makes longer iterations and convergence issues and it is eliminated by resilient backpropagation algorithm as stated by Riedmiller and Braun (1993). In this method, sign of the derivative alone is used to determine the direction of weight update and the magnitude is obtained by separate update function. The weight update is adaptive based on the rule: if the derivative is positive, then the weight is decreased by the update value and viceversa. If the weight change occurs in the same direction for several iterations, then the magnitude of the weight change is increased accordingly. This algorithm follows batch update.

8.5.6.3 LEVENBERG–MARQUARDT ALGORITHM

Levenberg–Marquardt (LM) algorithm computes the weight updation of the network using second derivatives based on Newton's method. According to More (1997), LM algorithm is less complex than gradient method, since the computation is made by Jacobian matrix instead of Hessian matrix. LM algorithm works as the combination of Newton's method and gradient method as follows. The weight update is done by:

$$w_{k+1} = w_k - \left[J^T J + \mu J \right]^{-1} J^T e, \tag{8.40}$$

where J is the Jacobian matrix which contains the first derivative of the error, I is the identity matrix, μ is the Marquardt parameter which is to be updated based on the decaying rate of the output, and e is the actual error as described in Coulibaly et al. (2000).

In eq 8.22, if the scalar μ is zero, this is Newton's method and if the μ is large, then this becomes gradient descent with small step change. Newton's method is faster and more accurate in optimization and hence the LM algorithm is more powerful than gradient descent.

8.5.7   DATA SERIES PARTITIONING

One typical method for training a network is to first partition the data series into three disjoint sets: the *training set*, the *validation set*, and the *test set*. The network is trained directly on the training set, its generalization ability is monitored on the validation set, and its ability to forecast is measured on the test set. A network's generalization ability indirectly measures how well the network can deal with unforeseen inputs, in other words, inputs on which it was not trained. A network that produces high forecasting error on unforeseen inputs, but low error on training inputs, is said to have overfit the training data. Overfitting occurs when the network is blindly trained to a minimum in the total squared error based on the training set. A network that has overfit the training data is said to have poor generalization ability.

8.5.8   WEIGHTS INITIALIZATION

The choice of initial weights is one of the important factors which influence much on the convergence of the network toward the global minima. The weights are initialized randomly for the NN in general. If the weights are very small, the net input will be very small or equal to zero and convergence becomes slower. On the other hand, if the weights are chosen largely, there is a possibility of saturation. In usual practice, the selection of random weights will be either in the range of −0.5 to 0.5 or −1 to 1. The other important procedure of choosing the weights is given by

Nguyen–Widrow (1990). In this method, the weights are initialized in the range of $-\beta$ to β, where β is a scaling factor defined as:

$$\beta = 0.7\sqrt[n]{m} \tag{8.41}$$

The network trained using Nguyen–Widrow weights provides improved training than that of the random weights.

8.5.9 RADIAL BASIS FUNCTION NETWORK

RBFN is a different feedforward NN model in which the activation function of the hidden unit is determined by the distance between input vector and a prototype vector.

RBFN training is done by two-stage procedure. In first stage, the parameters which govern the basis functions are determined by unsupervised methods and in the second stage the final-layer unit weights are determined. The basic architecture of the RBFN model is shown in Figure 8.9.

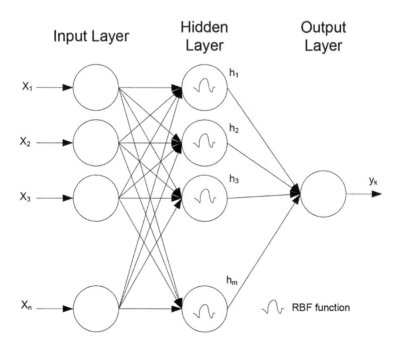

FIGURE 8.9 Radial basis function network architecture.

The RBFN method introduces a set of N basis functions, one for each data point in the form $\varphi\left(\left\|x - x^n\right\|\right)$, where φ (.) is some nonlinear function.

The nth function depends on distance $\left\|x - x^n\right\|$, usually measured as the Euclidean distance between x and x^n. The output of mapping is then taken as a linear combination of basis functions.

$$h(x) = \sum_n w_n \varphi\left(\left\|x - x^n\right\|\right)$$
(8.42)

In matrix form, $\varphi.w = t$, where $t = t^n$ and $w = w^n$. Here, t represents target vector and w represents linear weight vector.

$$w = \varphi^{-1}t,$$
(8.43)

where φ is a nonsingular matrix with distinct data points.

The basic function commonly considered is of Gaussian type.

$$\phi(x) = \exp\left(\frac{-x^2}{2\sigma^2}\right),$$
(8.44)

where σ is the smoothing parameter of the output.

Equation 8.36 is a localized basis function with the condition $\varphi \to 0$ as $|x| \to \infty$.

The basis functions are modified by Moody and Darken (1989) to smoothen the output. To smoothen the radial function, the numbers of radial functions are determined by the complexity of mapping rather than the size of data set.

Modified Gaussian radial function is given as:

$$\phi_j(x) = \exp\left(\frac{-\left\|x - \mu_j\right\|^2}{2\sigma_j^2}\right),$$
(8.45)

where x is the d dimensional input vector and μ_j is the vector determining center of function. The output layer is updated using eq 8.38.

$$y_k(x) = \sum_{j=1}^{m} w_{kj}\varphi_j(x) + w_{k0}$$
(8.46)

The most important advantage of RBFN in comparison with the MLP is choosing suitable parameters for hidden units without performing full nonlinear optimization of network. The formulation of RBFN by Gaussian hidden units has to be performed by unsupervised manner. To attain this, Haykin (2009) suggested methods such ask-means clustering and OLS.

8.5.9.1   K-MEANS CLUSTERING ALGORITHM

K-means clustering is the unsupervised learning algorithm which forms natural clusters of *K* numbers based on the similarities of the observation pairs. After forming the clusters, the average measure of dissimilarity of the observations from the cluster mean is minimized by using iterative descent algorithm. This algorithm is simple in implementation and computationally efficient. It assumes random choices of the cluster mean and chooses the smaller size *K*. If the number of clusters is chosen large enough, then the algorithm is capable of transforming the nonlinear patterns into linear patterns which can be propagated to the output linear layer.

8.5.9.2   OLS ALGORITHM

This algorithm starts by considering the network with one basis function. For each data point, we set the basis function center chosen randomly to input vector of that data point and then set the second-layer weight by pseudoinverse techniques. In subsequent steps, number of basis functions is increased incrementally. This is done by constructing a set of orthogonal vectors in the space *s* spanned by vectors of hidden unit and results in great reduction in residual sum-of-squares error. The algorithm is repeated until the residue error will be zero. Chen et al. (1991) suggested that the width of the radial function could be chosen in ad hoc manner.

RBFN plays similar role as that of MLP with few differences.

1) Hidden unit in MLP depends on weighted linear sum of inputs, whereas RBFN uses distance prototype.
2) MLP is more prone to local minima problem and convergence become slower. In contrast, RBFN with localized basis function takes only a few hidden units with faster convergence.

3) MLP uses many layers of hidden units and uses variety of activation functions, whereas RBFN uses single hidden layer with similar radial function.

All the parameters in MLP are determined at the same time using supervised training, whereas RBFN parameters are determined at two stages with unsupervised first stage and supervised second-stage computation.

8.6   RESULTS AND DISCUSSION

In this section, the forecasting of wind speed using the abovementioned stationary neural networks are studied in detail. The forecast methods are tested for the same set of sample data, measured at 15 min intervals which is utilized in the Chapter 3. In addition to the wind speed, the other contributing meteorological parameters such as wind direction, relative humidity, temperature, and barometric pressure are aggregated and considered as inputs for the model. These parameters have their influence in the wind speed based on the physical properties. The performance of the model based on the various metrics is analyzed. All the parameters having length of 300 samples are chosen to analyze the study as the interest of forecast zone is of short term only.

8.6.1   BPN-BASED WIND SPEED FORECAST MODEL

The wind speed forecast is obtained by developing appropriate BPN model. As already discussed in previous section, number of neurons in the hidden neuron is decided by trial and error approach. From the literature of Zhang et al. (1998), it has been concluded that ANN with single hidden layer is sufficient for forecast applications and hence the same is used for the models. Table 8.8 presents the results of various error criterions used for the BPN model with variation in number of hidden neurons. Figure 8.4 shows the variation in the mean square error for chosen BPN models and Figure 8.5 exhibits the regression value for the same models. From the results, it has been concluded that for 12 hidden neurons network model is able to produce good accuracy as the error criterions are comparatively less. The NN-based forecasting involves two steps, training and testing. During training, the historical time instant-based data which contains both the inputs and corresponding desired output, are presented to the network. In the testing process, the network maps the input with output by adjusting the weights and biases iteratively until acceptable output is met. This results in slow convergence, and it is usually based on the gradient descent optimization algorithm. To increase the convergence speed, various optimization algorithms as discussed in the previous section are considered and various forecast models are developed and tested for error performance.

TABLE 8.8   Performance Metrics Comparative Results for BPN Model.

Model	N (Hidden neurons)	R²	R	RMSE	MSE	MAPE	SSE
1	10	0.8953	0.9462	0.3806	0.1449	23.11	43.314
2	11	0.8405	0.9168	0.4628	0.2142	35.52	64.055
3	12	0.9071	0.9524	0.3539	0.1252	26.62	37.461
4	13	0.8930	0.9449	0.3795	0.144	28.72	43.064
5	14	0.8798	0.9475	0.4015	0.1612	29.92	48.212
6	15	0.8689	0.9321	0.4062	0.1649	30.14	49.456
7	20	0.8921	0.9242	0.3816	0.1456	27.92	43.526

Note: Figures in bold indicate better performing model.

The performance measures for different forecast models based on the learning methods are tabulated in Table 8.9. From the table, it can be concluded that the LM algorithm-based BPN provides good results in comparison with gradient descent momentum, scaled conjugate gradient, and resilient backpropagation-based models. The MSE and the regression values of the forecast models of the learning algorithms used in BPN models are shown in Figure 8.10. Hence, LM-based BPN model with 12 neurons is considered as the optimal forecast model. The regression parameter for the BPN model is shown in Figure 8.11. It is clear from the

TABLE 8.9   ANN Model Selection-based on Training Algorithms.

Model	Training algorithm	R	MSE	MAE	SSE
1	Gradient descent momentum	0.8511	0.3725	0.4632	111.39
2	Scaled conjugate gradient	0.7951	0.4999	0.5525	149.47
3	Resilient backpropagation	0.8791	0.3741	0.4673	111.864
4	Levenberg–Marquardt	0.9363	0.1658	0.3262	49.579

figure that the regression value of the model with 12 hidden neurons is 0.9524, which is comparatively higher than the other models with various hidden neurons considered here.

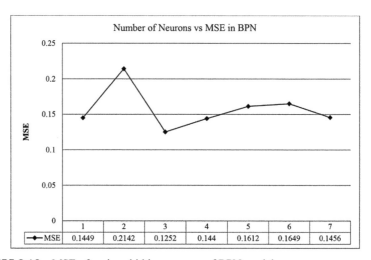

FIGURE 8.10 MSE of various hidden neurons of BPN model.

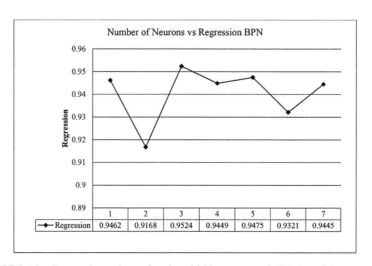

FIGURE 8.11 Regression values of various hidden neurons in BPN model.

The other parameter variations have been addressed by providing the different learning rate for the chosen model by providing different inputs between the limit of 0 and 1. As already discussed, it has been identified that the model with learning rate of 0.9 and momentum factor of 0.95 gives good accuracy in forecast results. The above-mentioned methodology is implemented using Matlab R2011b. The actual and forecasted results for the best BPN models are shown in Figure 8.12.

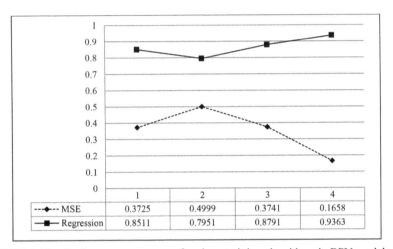

FIGURE 8.12 Performance measures of various training algorithms in BPN model.

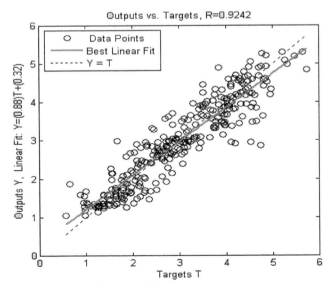

FIGURE 8.13 Regression of BPN model.

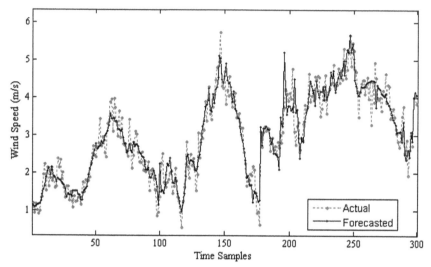

FIGURE 8.14 BPN output.

8.6.2 RBFN-BASED WIND SPEED FORECAST MODEL

The RBFN is trained by using the OLS learning algorithm. The effect of variation of spread factor β on required number of RBF centers is studied. The investigation reveals that for the spread factor of 0.07, the number of RBF centers reaches 50 with minimum performance measure. In addition to MSE other performance measures are also considered for all the neural network-based architectures and the obtained results are listed in Table 8.10. The forecasted output on comparison with the input wind speed obtained through simulation of RBFN is shown in Figure 8.15. On examining the results based on the performances, it may be concluded that the RBFN method provides the forecasts of wind speed for short-term period with better performance compared with BPN model.

Table 8.11 shows the comparison between the best chosen model of BPN (hidden neurons of 12, LM algorithm, learning rate of 0.9, and momentum factor of 0.95) and RBFN (spread factor of 0.07 with 50 hidden neurons). From the performance metrics it is clear that the RBFN model gives a good forecast comparatively, but there is no such drastic improvement. Moreover, it has been observed that though the RBFN is quick to train, it becomes slow for the test data and there exists close correlation between the experimental results between BPN and RBFN forecast models. More-over, selection of appropriate spread factor in RBFN is a challenging task. Hence, improvement in the BPN model will result in improved perfor-mance than choosing RBFN which is more complex.

TABLE 8.10   Performance Metrics Comparative Results for RBFN Model.

Spread factor	MSE	Regression (R)	Number of neurons
0.2	0.9966	0.9	5
0.15	0.9966	0.9275	10
0.1	0.1313	0.9487	25
0.09	0.1314	0.954	30
0.08	0.1314	0.959	40
0.07	**0.079**	**0.9649**	**50**

TABLE 8.11   Performance Comparisons of Best BPN and RBFN Model.

Performance index	Best BPN model	Best RBFN model
R	0.9524	0.9649
R^2	0.9070	0.9310
MSE	0.1252	0.079
RMSE	0.3538	0.2811

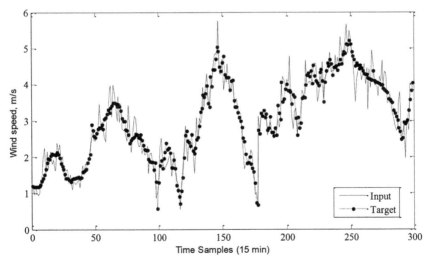

FIGURE 8.15   RBFN output.

8.7   CONCLUDING REMARKS

In this chapter various approaches to forecast the solar energy were discussed along with its merits and demerits. Also, various approaches to forecast the wind speed for wind energy conversion system using statistical methods and neural networks such as BPN and RBFN were discussed. Case study statistical models for wind speed forecast has been implemented and analyzed. BPN models were considered with various configurations and optimal model is chosen. The performance of training is improved by using RBFN model for forecast. The forecast accuracy based on the performance metrics has been compared among all the methods. This will provide an effective decision factor for the power producers in terms of reduced imbalance charges and penalties, period ahead energy market trading, and efficient wind plant construction, operation, and maintenance planning. These forecast models will help to improve the efficiency of the renewable power production and the similar models will be used for load forecasting and demand response of the grid to avoid the grid failure.

KEYWORDS

- **forecast models**
- **solar forecast**
- **wind forecast**
- **statistical methods**
- **numerical weather prediction**
- **neural network-based forecast models**

REFERENCES

Benmouiza, K.; Cheknane, A. Small-scale Solar Radiation Forecasting Using ARMA and Nonlinear Autoregressive Neural Network Models. *Theor. Appl. Climatol.* **2016**, *124*(3) 1–14.

Bishop, C. M. *Neural Networks for Pattern Recognition;* Claredon Press: Oxford, UK, 1995.

Box, P.; Jenkins, G. *Time Series Analysis: Forecasting and Control*; Holden-day Inc.: San Francisco, CA, 1976.

Box, G. E. P.; Jenkins, G. M.; Reinsel G. C. *Time Series Analysis: Forecasting and Control,* 4th ed.; John Wiley & Sons, Inc.: New York, 2008.

Brockwell, P. J. *Time Series: Theory and Methods;* Springer-Verlag: New York, 2002.

Brooks, C. *Introductory Econometrics for Finance;* Cambridge University Press: London, 2002.

Charalambous, C. In *Conjugate Gradient Algorithm for Efficient Training of Artificial Neural Networks;* Proceedings IEE-G, Circuits Devices and Systems, 1992; 139(3), 301–310.

Chatfield, C. *Time Series Forecasting;* Chapman and Hall/CRC Press: Washington, DC, 2001.

Chen, S.; Cowan, C. F.; Grant, P. M. Orthogonal Least Squares Learning Algorithm for Radial Basis Function Networks. *IEEE Trans. Neural Netw.* **1991,** 2(2), 302–309.

Cho, V. Time Series Data Forecasting. In *Encyclopedia of Data Warehousing and Mining;* Wang, J., Ed.; IGI Global, 2005; 1125–1130.

Chow, C.W., et al. Intra-hour Forecasting with Total Sky Imager at UC San Diego Solar Energy Testbed. *Sol. Energy* **2011,** 85(11), 967–977.

Coulibaly, P.; Anctil, F.; Bobee, B. Daily Reservoir Inflow Forecasting Using Artificial Neural Networks with Stopped Training Approach. *J. Hydrol.* **2000,** 230, 244–257.

Dong, Z., et al. Satellite Image Analysis and a Hybrid ESSS/ANN Model to Forecast Solar Irradiance in the Tropics. *Energy Conver. Manag.* **2014,** 79, 66–73.

Eftekharnjed, S.; Heydt, G. T.; Vittal, V. Optimal Generation Dispatch with High Penetration of Photovoltaic Generation. *IEEE Trans. Sustain. Energy* **2015,** 6(3), 1013–1020.

Fletcher, R. *Practical Methods of Optimization;* Wiley: New York, 1987.

Fukuoka, Y.; Matsuki, H. A Modified Back-propagation Method to Avoid Local Minima. *Neural Netw.* **1998,** 11, 1059–1072.

Haykin, S. *Neural Networks and Learning Machines;* PHI Learning: New Jersey, 2009.

Hebb, D. O. *The Organization of Behavior;* Wiley & Sons: New York, 1949.

IEGC. 2010. From Ministry of Power, Government of India: http://www.powermin.nic.in. (accessed March 1, 2013).

Kolmogorov, A. Three Approaches to the Quantitative Definition of Information. *Prob. Inf. Transm.* **1965,** 1(1), 3–11.

Li, Y.; Su, Y.; Shu, L. An ARMAX Model for Forecasting the Power Output of a Grid Connected Photovoltaic System. *Renew. Energy* **2014,** 66, 78–89.

Lorenz, E., et al. Irradiance Forecasting for the Power Prediction of Grid-connected Photovoltaic Systems. *IEEE J. Sel. Topics Appl. Earth Observ. Remote Sens.* **2009,** 2(1), 2–10.

Madsen, H., et al. Standardizing the Performance Evaluation of Short Term Wind Power Prediction Models. *Wind Energy* **2005,** 29(6), 475–489.

Malamatenios, C.; Choustoulakis, P.; Mengos, S. Wind Energy. In *Guidebook on the RES Power Generation Technologies;* Malamatenios, C., Ed., CRES, 2001; 1–7.

McCulloch, W. S.; Pitts,W. H. A Logical Calculus of the Ideas Immanent in Nervous Activity. *Bull. Math. Biophys.* **1943,** 7, 115–133.

Minsky, M. L; Papert, S. A. *Perceptrons;* MIT Press: Cambridge, MA, 1988.

Moller, F. G. A Scaled Conjugate Gradient Algorithm for Fast Supervised Learning. *Neural Netw.* **1993,** 6, 525–533.

Moody, J; Darken, C. Fast Learning in Networks of Locally-tuned Processing Units. *Neural Comput.* **1989,** 1(2), 281–294.

More, J. J. Numerical Analysis. In *Lecture Notes in Mathematics;* Watson, G. A., Ed.; Springer-Verlag, 1997; 105–116.

Nguyen, D.; Widrow, B. Improving the Learning Speed of Two-layer Neural Networks by Choosing Initial Values of the Adaptive Weights.*Proc. Int. Joint Conf. Neural Netw.* **1990,** 21–26.

Reikard, G. Predicting Solar Radiation at High Resolutions: A Comparison of Time Series Forecasts. *Sol. Energy* **2009,** *83*(3), 342–349.

Riedmiller, M.; Braun, H. A Direct Adaptive Method for Faster Backpropagation Learning: The RPROP Algorithm. *Proc. IEEE Int. Conf.Neural Netw. (ICCN)* **1993,** 586–591.

Rosenblatt, F. *Principles of Neurodynamics;* Spartan Books: Washington DC, 1962.

Rumelhart, D. E.; Hinton, G. E.; Williams, R. J. Learning Representations of Back Propagation Errors. *Nature* **1986,** *323*, 533–536.

Salomon, R.; Hemmen, J.L. Accelerating Backpropagation Through Dynamic Self-adaptation. *Neural Netw.* **1996,** *9*, 589–601.

Wagle, B. V. A Review of Two Statistical Aids in Forecasting. *J. R. Stat. Soc.: Series D (The Statistician)* **1965,** *15*(2), 191–196.

Wan, C., et al. Photovoltaic and Solar Power Forecasting for Smart Grid Energy Management. *CSEE J. Power Energy Syst.* **2015,** *1*(4), 38–46.

Zhang, G.; Patuwo, B. E.; Hu, M. Y. Forecasting with Artificial Neural Networks: The State of the Art. *Int. J. Forecast.* **1998,** *14*, 35–62.

CHAPTER 9

ENERGY MANAGEMENT SYSTEM IN SMART GRIDS

VIKRAM K. and SARAT KUMAR SAHOO*

School of Electrical Engineering, VIT University, Vellore, Tamil Nadu 632014, India

CONTENTS

ABSTRACT

The main aim of the electric utility companies is to ensure the continuous supply to the loads without any interruption. Utilities should also ensure the voltage levels, frequency, and reliability within the limits. In recent days, the modern research has introduced information and communication technologies for the power systems to act more intelligently. With the introduction of modern automation devices, two-way-communication systems and advanced control systems the traditional power systems are becoming smarter, thus transforming the traditional grid to "smart grid". At the distribution side, the smart meter integration and advanced metering infrastructure (AMI) is allowing the end user to monitor, control, and shift their energy usage according to their needs making "energy management" possible. The AMI also enables the utilities to make decisions based on the load analysis, identification, and location of faults, then allows reconfiguration of system strategies based on the conditions. This chapter presents an overview of the smart grid technologies, AMI, communication networks for smart grid applications, and cybersecurity challenges involved. The aim is not exhaustive rather it makes us understand how all the above technologies can make energy management system more realistic.

9.1   INTRODUCTION

In many countries, the present electrical and power systems are still synchronized with age-old electric grid and are planning for future advancements in the system. There is a need of moving forward for meeting future energy demand needs with a new kind of electric grid that not only assures the continuous power supply but also handles the increasing complexities, with the introduction of modern automation that strengthens present day electrical power systems (Amin & Stringer, 2008). The growing energy needs and essence of reducing the emission of carbon gasses with the integration of renewable energy resources have led the introduction of an intelligent electricity system that employs bidirectional flow of information for enhanced monitoring and precise controlling purposes called as "smart grid" (SG). With the advent of modern digital, electronic, and computerized equipment, the existing power systems is getting transformed to SG enabling for more reliability, security and cost-effective electricity to the consumers (Amin, 2011). Figure 9.1 shows the SG technologies.

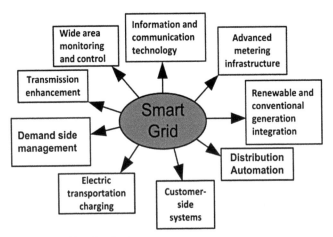

FIGURE 9.1 Smart grid technologies.

The power generation, transmission, and distribution are managed accordingly and intelligently based on the proactive scheduling of loads by SG. The SG acknowledges the supplier and end user about the energy usage behavior and also about different energy rates in accordance with the time, thereby providing an opportunity to the end user to manage the energy bills and even sell back to the grid, if surplus renewable energy is produced at home (Keyhani, 2011). SG supports the generation of electricity by the integration of conventional and nonconventional energy sources. The initiation of SG shall strengthen the overall reliability, operational performance, efficiency, and stability, and decreases carbon emissions for modern power systems (Jarrah et al., 2015)

The SG technologies have made the availability of real-time data to benefit the utilities for making more intelligent decisions during normal and hostile situations. With the increased monitoring, there shall be reduced failures, reduced maintenance, fewer outages costs, and assures the increased life of the power system assets. The advanced control technologies integrated with the advanced data communication technologies will reduce the power disorders and segregate faults and results in rapid restoration of outages (Gao, et al., 2012).

The SG enables wide area measurement system (WAMS) with the help of advanced communication system that can be utilized for controlling generating stations and planning for load scheduling consequently ensuring the stability and security. The power system stabilizer based on the WAMS technology is becoming popular. The communication setup plays an important role in coordinating generation plants, transmission system, system operator, power market, and distribution system (Wall et al., 2016). Table 9.1 shows the variations among traditional grid and SG.

TABLE 9.1   The Variations Among Traditional Grid and Smart Grid.

Traditional grid	Smart grid
End user informs when the power is disconnected.	Utility recognizes if power is switched-off and switches-on automatically.
Utility recompenses whatsoever it takes to meet peak demand.	Utility controls demand at peak. Reduces cost of utilization.
Wind- and solar-based generation has high penetration, hence difficult to manage.	The penetrations occurred by wind energy and solar energy causes no problem for load supply.
Distributed generation cannot be managed firmly.	Can handle distributed generation securely.
Ten percent of power loss in transmission and distribution systems.	The power loss lessened by 2% with the significant decrease in CO_2 emissions and minimizing customer bills.

An efficient transmission network plays a key role in carrying the power from generation stations to distribution stations. With the employment of communication network for transmission system can enhance monitoring in real time and can protect the system from the potential disturbances by reducing the losses and voltage variations thereby increasing the reliability and ensures the optimal utilization of transmission network (Nafi et al., 2016). The automation of substation and distribution will be the important enablers for the smart distribution systems. The increased usage of distributed energy resources (DERs) will be a significant feature of forthcoming distribution systems. SG switches the peak loads by identification and establishing different DERs (Yu et al., 2015).

The communication infrastructure employed will act an important role in interchanging information between the distribution management system (DMS) and substations. The main work of the distribution system substation operator is to regulate the DERs in an organized way for improving the reliability and power quality of the distribution system. In the smart distribution system, the exchange of information is expected between end user and the distribution system substation operator, for increasing the stability.

This accelerates end user-based power generation using solar panels and storage devices that are not only connected with the grid but also enables the two-way flow of electricity and communications is an important feature of the smart distribution system. This enables end user to make the trade with the utilities if surplus energy is generated or if the tariff of generation is high compared to receiving (Meliopoulos et al., 2013).

The power industry considers demand response (DR) as a valuable resource through which the standardization of modern grid is possible. One of the foremost objectives of SG is to develop the modern techniques and technologies that make use of DR for optimizing the dynamic performance of grid with the participation of end user. The DR is a significant resource through which the power demand shall be dynamically and resourcefully adjusted with generation side resources that lessen per-unit-price which is based on the demand fluctuations. DR provides a chance to end users to play an important role in the operation of SG by shifting or decreasing their power utilization during peak timings in response to the timely based tariff. By peak-load shifting, decreasing the usage of power when less generation is expected and by the emergency reaction, the end user is going to act a dynamic role in the improved operation of the distribution system (Haider et al., 2016).

The SG-based end user can make use of the updated data displayed by power companies and can define his/her electricity usage behavior by shifting smartly during the peak loads by using home area networks (HANs) to reduce the electricity bills and thus enhancing the reliability of power delivery systems in an intellectual way. Thus for enabling the SG, end user requires appropriate communication technology to interact with the utility operators, distribution systems, and the electricity market (Mahmood et al., 2015).

The smartness of SG lies in the integration of extensive range of physical power assets and information resources for advanced monitoring and control, making distribution infrastructure more intelligent by employing multiple advanced technologies such as building energy management systems (BEMS), home energy management systems (HEMS), smart-meter data management system (MDMS), distribution automation (DA), DMS, energy management systems (EMS), and advanced metering infrastructure (AMI) for scheduling generation and transmission systems enabling SG operations for real-time information exchange with power markets allowing for power trade-off and scheduling. The utilities must work together with different service providers for guaranteeing the systematic functioning of the SG. Data exchange with the end users is significant for the utility to implement the DMS (Bayindir et al., 2016).

Most suitable regulatory policies must be framed for the smooth unification of the various technologies, including the energy storage and DER aggregators into the SG market. The per-unit cost information should be updated online for every shorter interval (hours or even minutes). The encouraging areas for the service in SG includes anticipating the renewable generation, planning tariff, end user complaint management, HEMS, setting up and commissioning services, financial management, etc. (Shomali & Pinkse et al., 2016).

The SG is employing various kinds of advanced sensors, automatic controls, and makes use of advanced software by utilizing the actual data to identify and segregate the faults and to rearrange the distribution network to the impact on the end users and thus becoming itself as a "self-healing grid." The foremost objectives of self-healing grid is to enhance the overall reliability and stability of the distribution network, and this can be carried out by the reconfiguration of the reclosers, switches, and relays installed on the distribution feeder that instantly segregates the faulted section of the feeder and establishes service again to as many end users as possible from alternative resources or feeders (Emmanuel & Rayudu, 2016).

It is very important to take the decisions very quickly in order to implement the control actions that are to be implemented within the specified time defined by SG committee (generally ranges from few milliseconds to less than 5 min); hence, it requires standard communication technology for reconfiguring the system. To achieve this, it requires a standard communication with high bandwidth as defined by smart grid interoperability panel (SGIP) based on application, for example, AMI requires a bandwidth of −100 kbps per node/device with an optimal latency between 2 and 15 s in home applications (generally uses Zigbee or power line communications (PLC) communication) and the delay more than this shall affect the DR. The data that is typically aggregated at AMI (aggregation point) is to be communicated to utility center shall demand a bandwidth requirement of about 500 kbps. The latency requirements of DR can be estimated from as small as 500 ms up to 2 s to minutes. For actual monitoring and control, latency requirements are very low. According to Alcatel, the maximum latency requirement for monitoring is 20 ms even though companies such as UTC and Avista defines that it is below 200 ms. According to Avista and UTC, the data rate necessary for DER will be 9.6–56 kbps same as AMI. (Mets et al., 2014).

SG is very important for government and end users because it is improving the quality of power. But dependency of SG on computer networks and internet resources is making it more viable to security attacks introducing the privacy problems such as data protection of end user information. The cyberattacks such as Stuxnet, Night Dragon, and Duqu were discovered in the applications of SG. Thus policies and regulations should be made for promoting the best practices and should plan that utility companies strictly adhere to security as an important factor and should consider from a holistic point of view (Wang & Lu, 2013).

International Electrotechnical Commission (IEC) is responsible for standardization of electrical and electronics related fields worldwide. The standards in Figure 9.2 shows core standards for SG architecture directed by IEC technical committee 57. The IEC TC57 commission is employed for defining standards for electric power system management and related data communication for the real-time operations and planning of generation, transmission and distribution, and their respective information exchange so as to encourage power market. Table 9.2 gives information about different SG technology areas and their respective hardware and software systems (IEC, 2010).

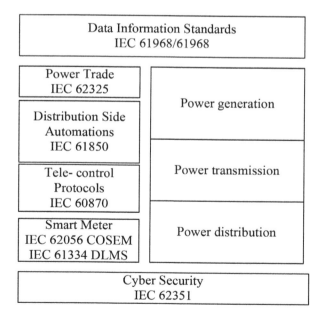

FIGURE 9.2   Fundamental benchmarks for smart grid, IEC TC57 recommended architecture.

TABLE 9.2 Advanced Technologies for Implementing SG.

Technology area	Hardware	System and software
Wide area monitoring, planning and control	PMU (phasor measurement units).	SCADA, WAMS, WAAPCA
ICT (information and communication technology)	Communication equipment (broadband wired and wireless access) .	CIS (customer information system), ERP (enterprise resources planning software).
Renewable energy and distribution generation integration	Battery storage, converter and inverter, Smart control systems, communication devices.	SCADA, DMS, geographic information system (GIS).
Transmission standardization	FACTS, PMU, Synchrophasors, connected with communication devices.	Automated recovery systems.
End user side management	Remote controlled distribution generation and storage, advance sensors, actuators and transducers.	Outage management system (OMS), GIS, DMS.
AMI	Smart meter, sensors, Actuators, smart displays, home gate ways.	MDMS
Electric vehicle charging infrastructure	Batteries storage, converters, inverters, smart switches.	Smart power billing, G2V, and V2G methodologies.

9.2 ADVANCED METERING INFRASTRUCTURE

SG is the combination of many intelligent subsystems having their importance for strengthening the overall performance of grid. AMI is one of the most significant and an essential technology, holding the responsibility for the collection of data and information from different end user loads and analyses it for utility centers. AMI has become most important tool for implementing the control and command signals for demand-side management (DSM) aspects by end users and utility centers. This chapter explains the relationship between the SG and AMI enlightens all the important areas of the AMI (Anda & Temmen, 2014).

AMI allows two-way communication between end users and utility companies with the integration of smart meters, communication networks, and data reception and management systems as shown in Figure 9.3. The important aspects of AMI are the remote location smart meter reading without any errors, identification of problems in the communication network, reporting and analyzing the load usage, reviewing the energy usage and looking for alternative supply instead of completely load shedding.

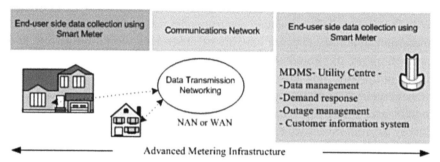

FIGURE 9.3   Advanced metering infrastructure block diagram.

9.2.1   SUBSYSTEMS OF AMI

AMI is a combination of software and hardware codesign that perform a vital role in transmission, distribution, and measurement of energy consumption for utility centers. The various subsystems of AMI include: (Mohassel et al., 2014)

A. Smart meters: The end users energy information is collected at regular time intervals and is transmitted through communication networks to the utility and in return command signals and pricing information from the utility is conveyed to the end user.

B. Communication networks: The advanced data communication for AMI involves in the bidirectional flow of information from end user to utility and vice versa. For this purpose, networks such as PLC, broadband over power line (BPL), fiber optic communications, public networks, or fixed radio frequency networks can be used.

C. Meter data acquisition system: The smart meters from individual users in an area transmits the information, to data concentrator units that receive or sends data collected (or command signals) to and from neighborhood area network (NAN) or HAN to utility (MDMS).

D. MDMS: The utility center collects stores and evaluates the metering information.

9.2.1.1 SMART DEVICES

With the advent of the advanced electronic devices such as modern sensors and smart actuators that are capable of measuring, communicating, and controlling are deployed at the end user premises according to the usage and functionality has instituted smart homes that deal with the energy management through the smart meters. Two important functions of smart meters are: metrology (hardware to measure and control the flow of energy to/from building) and secondly is communication (Uribe-Pérez et al., 2016). Every smart meter has to provide the following functionalities:

- *Quantitative measurement:* The accurate measurement of data is very important for AMI, the HAN generally depends on Zigbee and PLC for communication of measured data from sensors to the smart meter and vice versa. Hence, smart meter should precisely assess and quantify the different topologies, principles, and methods.
- *Control and calibration:* A smart meter needs the standardization owing to dissimilarity in voltage values, sensor tolerances or other system errors, and recompenses the small variations in the system. Smart meters also deliver remote calibration and control ability through communication links.
- *Communications:* The information collected by the smart meter can be communicated to the utility via the wired or wireless connection. It should also update the information to end user or receive the commands from utility and can only be possible through standard communication.
- *Power management:* Generally, the sensor-based appliances communicate to the smart meter through the sensor networks, so the smart meter should manage the nonelectric metering appliances based on the priorities, whose power management is important for maximizing the service of battery and that to enhance the service of the network.
- *Display:* The DSM starts from the customer response and participation. For enhancing the quality of AMI system, the information

supplied from utility should be easily acknowledged to the end user through smart meter display.

- *Synchronization:* Timing synchronization is significant for the consistent communication of information to the utility to support functions such as evaluation of data and accurate billing. This is mostly desirable in a wireless network that has an asynchronous communication protocol.

9.2.1.2 COMMUNICATION

The smart meter is a sensitive and complex device that handles large data between home appliances and utility center without any disruptions. The smart meter data is most trustworthy and the access is limited to few people. The communication standards and strategies are framed to safeguard the data transfer within the network and should be protected. Every smart meter is assigned with a unique identity and also all the appliances associated are also assigned with similar identity so as to secure the cryptographic encryption. The communication network should also support the smart meter even if power outage happens. Communication technologies employed should be economic, should have better transmission ranges, with standard security features, and should provide the required bandwidth (Ma et al., 2013).

The end user premises enabled by HAN mainly implements AMI and DR. For coordinating smart meter for its monitoring purposes, HAN deploys various wireless technologies such as Bluetooth, WiFi (IEEE 802.11), and Zigbee (IEEE 802.15.4). Wired solutions may include the usage of Ethernet and PLC. Though wired communication supports good data rates and security, Ethernet involves high cabling costs and less flexibility compared to wireless. The usage of PLC for HAN is still in preliminary stages (Han & Lim, 2010).

NAN or field area networks (FAN) are mainly employed between HAN and wide area networks (WAN). There are two IEEE standards that are carefully related with NANs. The IEEE standard 802.15.4G mainly deals with an out-of-door environment with relatively low data rates (~less than 100 kbps) and associated with wireless smart metering utility network (SUN). Second, IEEE 802.11 s is closely related to the network operations such as node delivery and route selection of SG NANs. The privacy of data must be ensured from cyberattacks for SG NANs (Meng et al., 2014). Table 9.3 shows important communication technologies for AMI.

TABLE 9.3   Important Communication Technologies Used for AMI.

Technology	Home area networks	Neighborhood area networks	Wide area networks
Wired Technologies	• *Ethernet* • *High data-rate PLC* (multimedia, electric vehicle) • *Low data-rate PLC* (home automation, electric vehicle) • *Copper pair*	• *Low data-rate PLC* (Lon works, G1, G3…) • *Fiber* (FTTH) • *DSL* • *TV cable* • *Phone network*	• *PLC* • *Fiber* • *DSL*
Wireless Technologies	• *WiFi* • *Home automation standards* (Zigbee, Z-Wave, etc.) • *Energy and Metering Standards* (M-Bus, KNX) • *DECT*	• *RF* (point to point) • *RF Mesh* • *Cellular* (GPRS, 3G, LTE) • *WiFi* • *Wi-Max* • *Satellite*	• *Cellular* (GPRS, 3G, 4G/LTE) • *Wi-Max* • *Satellite*

WANs serve for SG between the NAN and utility center. WAN employs a high-bandwidth network for providing backhaul communication between different substations, distributed automation, and data aggregation points covering for thousands of kilometers apart. Reliability and security are the most important aspects of the WANs. Most of the utility operators such as AT & T, Verizon, and Sprint shall make use of private WANs for increased security instead of depending on public networks (Ho et al., 2014).

9.2.1.3   METER DATA ACQUIREMENT SYSTEM

The meter data acquirement system (MDAS) mainly deals with the data acquirement from automatic meter reading (AMR) at end user premises and within distribution systems for increased monitoring and for better planning of the decisions in order to reduce the losses and faults (Pathak, 2013).

Important features of MDAS include:

- collection of data from the individual user and communicating to corresponding area data concentrator units (DCU) for decision making,
- issuing notifications and generation of alarms when needed,
- preparation of reports based on the energy usage,
- increasing, ease of usage and monitoring for end users and utility centers.

Challenges faced by MDAS include:

- Improper hardware connections make systems weak. Hence connection issues should be rectified for proper recording of data and then communicating. Issues of smart meters and modems should be resolved.
- MDAS software should analyze the load usage and then it should be communicated to the server at the specific timing every day. If reception of data is not recorded by the server for 2–3 days continuously then it must be attended manually for rectifying issues.
- Network issues must be tested and resolved immediately for better data communication.

9.2.1.4   SMART-METER DATA MANAGEMENT SYSTEM

MDMS is a top priority area for system planning and decision making by SG and supporting it to become self-healing and resilient. MDMS is a combined system that is interconnecting the related data from different sources, becoming a standard centralized system for utility centers for their operations and management.

9.2.1.4.1   Features of MDMS

Data collection and synchronization: MDMS plans, operates, and schedules the data in an appropriate context for utility center. Standards based on the interfaces facilitate data to be consumed by MDMS from smart meter systems or SG devices.

- *Distribution network and power quality:* MDMS is responsible for load modeling and shall uphold the network connectivity maintaining the power quality and reliability.
- *Validation, estimation editing:* Servers are provided with high logical analysis for the data handling and data management.
- *Exception management:* MDMS manages exclusions and inclusions of connections based on fault analysis and also schedules the remedies based on the requirements.
- *Billing extracts:* MDMS plans billing exactly according to the load usage and shall be framing the cost analysis structure accordingly.
- *Analytics and reports:* The analytics and reports of MDMS shall correlate the data analysis to become utility business model.
- *Virtual and net metering:* A virtual meter is a careful combination of some subgroup of the smart meters in an area. Segmentation of the whole area under NAN is divided into small clusters. Each cluster will have virtual data aggregator that manages end user premises smart meter for notices, alarm's, and analysis so as to manage as if a single physical meter.

9.2.2   ADVANTAGES OF AMI TECHNOLOGY FOR SG

There many operational benefits with AMI that makes SG interoperable,
- *Functioning advantages*—The accuracy of metering reading is improved. The energy theft can be identified easily. The power outages and restorations can be managed easily.
- *Commercial advantages*—AMI brings the commercial advantages to utility centers by reducing personnel with automatic outage management and by reduced maintenance costs.
- *End user benefits*—AMI coordinates end user in the management of load based on tariff analysis, helps in reducing bills. Thus, end user management will also improve the flexibility and reliability of the SG.

9.2.3   CHALLENGES OF AMI

Even though AMI has many benefits to society, there are three major issues for its implementation and establishment that include the high cost of initial investment, interoperability of many technologies working together, and standardization.

- *High cost of investment:* AMI is the workforce in the implementation of SG. It is the combination of many software and hardware systems. So initially for the establishment of all these advanced technologies along a wider area requires high capital investment. Also for maintenance, it requires high-skilled personnel.
- *Interoperability:* Advanced technologies such as AMI are always complex. AMI integrates many advanced technologies such as communication networks (WAN, NAN, and HAN), data management systems, outage management system (OMS), DA, DER, smart metering, geographical information systems (GIS). All these technologies should coordinate each other and should work with interoperability.
- *Standardization:* Institutions such as National Institute of Standards and Technologies (NIST) has set Priority Action Plan (PAP) committees to define the standards for various technologies involving in deployment of SG, IEC, Institute of Electrical and Electronics Engineers (IEEE), have defined interoperability standards for uniform and successful deployment, there is a need for careful study and research on these standards while deployment of AMI technology.

9.3 COMMUNICATION NETWORKS PREREQUISITES FOR SG APPLICATIONS

One of the prime concerns for the SG is to develop the interoperable standards so as to make different technologies such as power engineering, communications, information technologies, and control engineering to come together for achieving required goals and standards (NIST, 2010). Interoperability is a unification of many technologies that includes communication networks, computing systems, intelligent electronic devices, smart machines, and their applications that have the capability for interchanging data with utmost security among each other. The SG applications include different network requirements, in terms of bandwidth, latency, data compression, and congestion management.

The interoperability architectural perspective (IAP) has three significant standpoints from the point of architectural: (1) Traditional power systems, (2) Advanced data communications, and (3) Smart information technology. SG interoperability can be achieved by considering these three standpoints. The aim of every standpoint's design is to adapt interoperability between the applications of the SG. All the above three technologies are responsible for all logical and functional considerations for SG and there is a need to address them with respect to their architecture and relatively should aim to derive specific relation with SG applications (IEEE Std. 2030, 2011).

9.3.1 COMMUNICATION NETWORKS ARCHITECTURE FOR SG

Based on the type of SG applications, the communication strategies are broadly classified into HAN, NAN, and WAN; these three networks are responsible for managing the entire appliances and applications from centralized utility center (Xi et al., 2012). Figure 9.4 below shows a network model with different networks for SG.

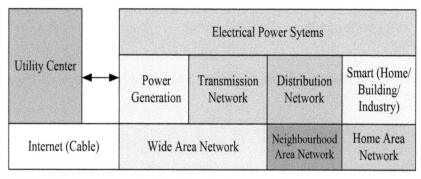

FIGURE 9.4 Network model with different networks for SG.

9.3.2 HANS FOR SG

The HAN is the most important technology for SG that enables bidirectional data communication for managing DR by utilities. The most important applications for HAN are home automation, building automation, and industrial automation. Authorizing and control are two necessary functionalities in the HAN. Authorizing is quantified to recognize and manage different appliances that form a self-organizing network. Control is an important functionality for safeguarding interoperability within the SG. The HAN consists of the smart meter, home gateway (HGW), intelligent electronic devices, smart sensors, actuators and smart appliances. HAN generally employees a star topology with either wired technologies (e.g., Ethernet and PLC) or different wireless technologies (e.g., Zigbee and WiFi) (Yu et al., 2011).

The relationship between the SG (utility center) and the end user is very important. The conceptual reference model for HAN illustrates two different components working for the end user, first, the smart meter, and second, the ESI (energy service interface), which operates as the gateway to the end user premises network. The above components analyze the electricity usage, stores, and communicate the data to utility center to perform service- and maintenance-related operations and tariff- and DR-based operations. (Hossain et al., 2012). Table 9.4 shows the comparison of HAN communication technologies.

TABLE 9.4   Comparison of HAN Communication Technologies.

Technology	Standardization	Operating frequency	Range (m)	Security	Cost	Data rate
Ethernet	IEEE 802.3	125 MHz	100	High	High	10 Mbps–1 GbpS
PLC	IEEE 1901	2–100 MHz	10–100	High	Medium	10 kbps –200 Mbps
Bluetooth	IEEE 802.15.1	2.4 GHz	10–50	128-bit E0	Low	0.7–2.1 kbps
Low Power WiFi	IEEE 802.11	2.4 GHz	30–100	WPA2 TLS/SSL	Medium	5–100 kbps
Zigbee	IEEE 802.15.4	2.4 GHz	10–75	128 bit AES	Low	250 kbps
Z-wave	G.9959	868.42 MHz or 908.42 MHz	10–30	TDES	Low	9.6–40 kbps

The SG implements energy efficiency and DR to increase its worth as an established and long-lasting infrastructure investment and assures return on investment in the short term. The main aim of HAN is remote monitoring and control of the electrical appliances such as thermostats, air conditioning (AC), vehicle charging, etc. The smart meter has an ability for connecting all the home applications using wireless connection using Zigbee or Wi-Fi that operate under the same frequency (usually unlicensed 2.4 GHz). The smart meters can control all the home appliances and prepares a comprehensive data on the power utilization of each appliance (Mouftah & Erol-Kantarci, 2016). Figure 9.5 shows different routing protocols for HAN.

Routing protocols for HAN

Routing for Wireless HANs

Vikram, K et al.: A survey on wireless sensor networks using Zigbee communication for HAN is carried.

C. Gomez et al.: A survey of wireless home automation technologies and their respective architectures was carr

J. Song et al.: Application oriented wireless mesh network communication protocol for process automation applications (Wirelesshart) for industrial control.Z.

Jindong et al. : An industrial wireless mesh network Elhfr based on graph routing.G. Mikhail et al., The z-wave

S. Petersen et al.: Wirelesshart versus isa100.11a for industry applications

D.-M. Han et al.: Smart HEMS (home energy management systems) using IEEE 802.15.4.

Routing for PLC HANs

J. Heo. et al. : HAN using PLC with Adaptive channel state routing

P. Darbee: INSTEON

Routing for hybrid HANs

C. Jin et al. : Smart HAN with wireless communication and PLC.

IPV6 RF-PLC Routing

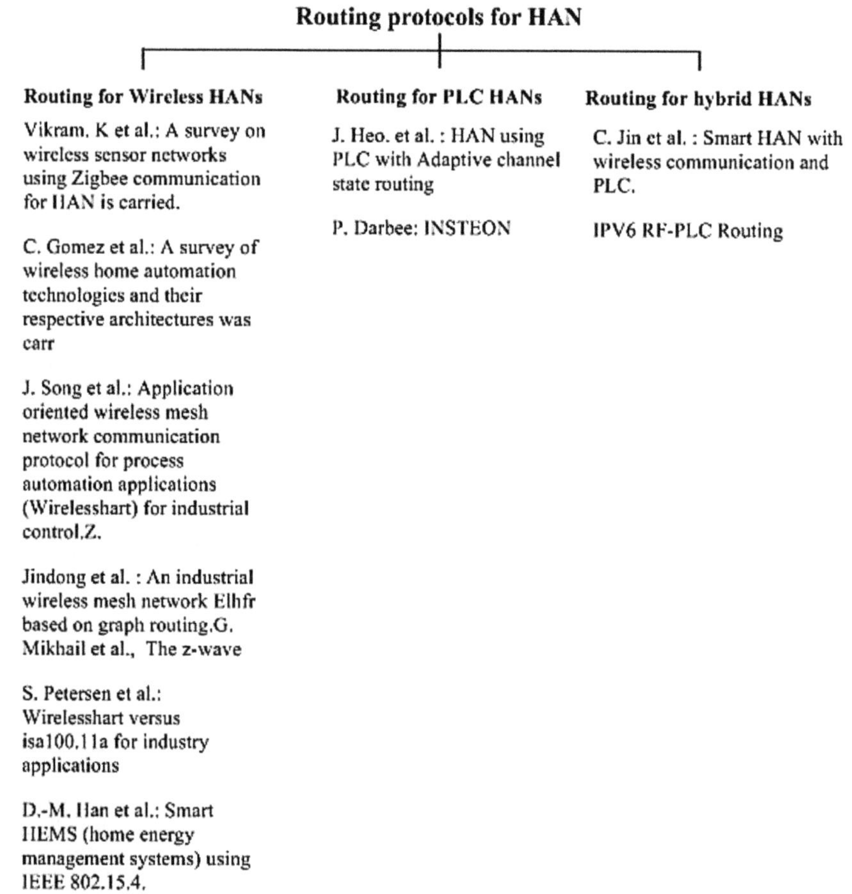

FIGURE 9.5   Routing protocols for HAN.

9.3.2.1 IMPORTANT SECTIONS OF HAN

Some important examples of DSM applications are:

- social energy efficiency,
- intelligence enabled power tariff, and
- efficient demand control.

9.3.2.2 ADVANTAGES OF HAN

- HAN permits the end users and lets the SG infrastructure to be benefited by the end users openly; this involvement of end users will benefit the utilities to manage peak loads.
- Utilities are informed using HAN about the electricity usage of every individual end user and provides centralized access for utility centers to control all the appliances at the end user premises.
- The main aim of HAN is to make the SG hassle-free by controlling or shifting the loads so as to save from potential blackouts.
- HAN enables end users to control their energy bills by shifting their loads from peak timing to normal load timings.

9.3.2.3 CHALLENGES OF HAN

- The integration of various technologies such as automation, wireless connections, and security is a challenging task.
- Interoperability of all the technologies for a common cause such as home energy management is an important concern and modern solutions should be acceptable by the market.
- The end user confidentiality and data security is a prime issue that needs to be addressed.

9.3.3 NEIGHBORHOOD AREA NETWORKS

The NAN shall connect the intelligent electronic devices such as smart meters to the AMI. NAN can also be termed as FAN. NAN plays a major role in connecting the distribution side appliances. The NAN covers an area of 1–10 km, working with the data rates between 10 and 100 kbps. NAN is operated by various technologies such as on the wireless side WiFi, radio frequency (RF) technologies, Worldwide Interoperability for Microwave Access (WiMAX), cellular (3G and 4G), and Long Term Evaluation (LTE) and on the wired side, PLC, Ethernet, Data Over Cable Service Interface Specification (DOCSIS) are promising preferences to use (Saputro et al., 2012).

The most important applications of NAN are as enlisted below, meter reading, DA, DR, prepaid payment, electric transmission and distribution monitoring, utility updates, program and configuration updates, outage resource management, time-of-use (TOU) pricing, service-based switching operation, end user information and message alerts, buildings network admin, etc. (Kuzlu et al., 2014).

Meter reading mainly collects information from appliances about their energy usage and communicates it to the utility centers using bidirectional communications. Smart meter readings strengthen better energy usage management by monitoring the power usage of every individual end user from utility center. Some important applications such as on-demand meter readings allow the end user to know about their energy usage queries when required immediately. Smart meter interval is software-based utility application that collects information from smart meters at scheduled intervals several times a day by utility centers.

Broadcasting of price information to user is one of the most important aspects of TOU that allows the end user to schedule their loads accordingly to the price information. Real-time pricing (RTP) offers information about short-term varying price information (e.g., variation of price in 10 min or 30 min) that increases because of sudden variations in load. Critical peak pricing (CPP) shall inform users about the pricing at very high peak demand.

DR is a very important in SG tasks. DR allows utility centers to control load at the end users premises such as controlling of thermostats, electrical vehicle charging, air conditioning, etc. in relative to peak timings. DA allows utilities to monitor all the important services and operations such as connection or disconnections to improve the reliability of SG.

Outage and restoration management (ORM) is a significant operation of SG that detects outages of power immediately. The problems of low voltages and high voltages can be learned immediately from smart meters readings. The problems involved in electrical vehicle charging can be resolved. The ORM makes SG more intelligent and stabilizes the performance by better utility management. Firmware updates shall update the background software that runs SG and fixes the bugs for improving performance. Customer information and messaging information shall allow the end user to know about their usage immediately.

There are network standards, such as WiFi (IEEE 802.11), Zigbee alliance (IEEE 802.15.4), Wi-MAX (IEEE 802.16), and WSNs may be used to implement NAN's for SG communication. There are two important latest IEEE standards that are most relative for SG-based NAN's (Meng et al., 2014). First, IEEE 802.15.4g insists on Physical Layer (PHY) and Medium Access Control (MAC) layer architecture of SG communication networks, Second, IEEE 802.11s addresses network operation issues of SG. The IEEE 802.15.4g targets mainly low data rate wireless communication in the outdoor environment and wireless SUN necessities. The SUN mainly aims for the very large dispersed network that operates with low power requirements. Smart Utility Network (IEEE 802.15.4g) (SUN) contains the large number of wireless devices that widespread over a large area and operates with efficient routing algorithms for data communication (IEEE Std., P802.15.4g, 2011). SUN operates in unlicensed frequency bands (2.4 GHz) and has to withstand interference with another wireless communication system (IEEE 802.11) operating in the same frequency band. (Vikram & Venkata Lakshmi Narayana, 2016). This standard has a lot of scope for future research. Figure 9.6 represents the routing protocols for NAN.

Routing protocols for NAN

QoS routing for Wireless NANs	Reliable routing for PLC NANs	Reliable routing for wireless NANs	Secure routing for Wireless NANs
Vikram. K et al.: A survey on wireless sensor networks using Zigbee communication for HAN is carried	S. Liang et al.: A communication network for multipath routing using broadcasting algorithm for PLC.	T. Iwao et al.: Wireless mesh networks for dynamic data forwarding	Li et al.: SG using homomorphic encryption for secured information.
	W. Gao et al. : A PLC network with improved routing based on AODV.	S. D. Haggerty et al.: HYDRO	Bartoli et al.: SG m2m networks for secured lossless aggregation.
	M. Biagi et al. : Location based PLC for SG.	J.-S. Jung et al.: Improving IEEE 802.11s reliable routing for SG.	
		H. Gharavi et al.: Multiple gate based SG communication network.	
		D. Wang et al: AMI for SG using routing protocol and low power and line lossy networks.	

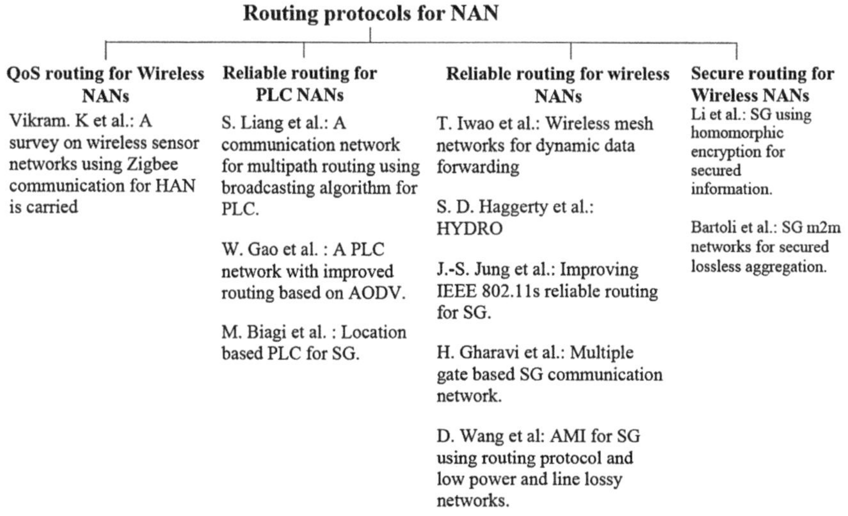

FIGURE 9.6 Routing protocols for NAN.

The IEEE 802.11s is an improvement for the existing protocol IEEE 802.11 in terms of better packet delivery ratio and route selection for multihop networks by improving the RF parameters at MAC layer. This protocol also includes the features of on-demand routing protocol and tree-based proactive routing protocol. IEEE 802.11s offers high reliability and high-speed data transmission with better routing for wireless NAN applications (Jung et al., 2011).

9.3.4 WIDE AREA NETWORKING

WAN offers communications linkage between SG applications and utility systems. WAN include two types of networks: core network and backhaul communication network. The core network connects a metropolitan network of the utility and substations. The backhaul network connects NAN (data aggregation points) to the core network. The WAN is a very large network covering thousands of square miles with the data transmission ranges upto 10–100 Mbps. In most of the cases, communication technologies used for WAN operations are public networks such as wired broadband lines or cellular networks (4G or 3G) (Terzija et al., 2011). But in recent times, security issues are questioning the viability of the public networks and to address the issues of WAN, particularly data security, private networks-based communication is adapted or thoughtful in the recent times. For reducing the cost of infrastructure, a virtual private network (VPN) is emerging which is a combination of both public and private network but with special traffic segmentation and including security, features that make VPN similar to the private network (Trilliant, 2010). Table 9.5 shows traffic and required quality of service for SG technologies.

The applications of WAN include wide-area monitoring, protection, and control (WAMPAC) most important next-generation-based solutions. The above strategies improve the power system planning and operations and enhance the reliability of the monitoring techniques (Juanjuan et al., 2011). The infrastructure used for establishing wireless WAN includes similar protocols such as WiMAX, 3GPP, and RF mesh are used for back-haul network and can be considered as part of the NANs. For wired options, passive optical networks (PONs) and Digital subscriber line (DSL) or can be employed. The metro Ethernet can be employed for the core network with some wired protocol such as internet protocol/multi-protocol label switching and SONET-fiber.

TABLE 9.5 Traffic and Required Quality of Services.

Technology	Traffic Types	Description	Bandwidth	Latency
Advanced metering infrastructure (AMI)	Meter reading	Energy consumption is reported using smart-meter (Ex: for every 15 min meter reads and stores and usually communicated for every 4 h).	10 kbps.	2–10 s.
	DR (demand response)	During peaks utility shall inform the end user for decreasing or shifting the loads.	Low.	500 ms ~ 1 min.
	Connects and disconnects	Connects/disconnects customer to/from grid.	Low.	A few 100 ms to few minutes.
Substation	Phasor measurement unit (PMU)	The most important power flow measuring technologies deployed for wide area situational awareness (WASA).	A few 100 kbps.	20 ms–200 ms
	Substation SCADA	4 S interval polling by master to all the intelligent electronic devices in the substation.	10–30 kbps	12 ms~20 ms
	SMART substation	Major advanced technologies such as DER might use GOOSE communication outside substation.	–	12 ms~20 ms
	Surveillance	Video side surveillance for rectifying damages due to thundering effects.	A few Mbps	A few Seconds.
		For controlling/protecting/restoring circuits.		
Distribution network	Identifying fault location, isolation and restoration.	Power quality and Volt/Var optimization on distribution systems.	10–30 kbps	A few 100 ms.
	Power system optimization.	Video and voice access to field personnel.	2~ 5 Mbps.	25~100 ms

TABLE 9.5 *(Continued)*

Technology	Traffic Types	Description	Bandwidth	Latency
	Workforce access.	For prediction and Proactively gathering and analyzing nonoperational data for potential asset failures.	250 kbps	150 ms
	Resource management	Isolation of circuits in response to faults indeed protecting power equipment.	–	–
Microgrid	Protection	Control and monitor the operations of the whole micro grid for energy management with SG.	–	100 ms~10 s
	Operation optimization		–	100 ms~1 min

9.4 CYBERSECURITY CHALLENGES FOR SG

The SG relies on superior computing and intelligent networking technologies for improving the reliability of power systems. For achieving this, many intelligent electronic devices are interconnected to the communication network, raising the issue for cybersecurity, and should be treated as a high priority issue by SG community (Ericsson, 2010).

Cybersecurity is defined for specific applications and domains of SG. Some standards address the operator, while others contain very detailed implementation requirements. Table 9.6 shows the cybersecurity standards for SG (European Commission, 2012):

TABLE 9.6 Cybersecurity Standards for SG.

Standard	Description
IEC 62351-1 to 6	Power systems management and associated information exchange—Data and communications security.
NERC CIP-002 and CIP-003 to CIP-009	NERC Standards CIP-002 through CIP-009 provide a cybersecurity framework for the identification and protection of Critical Cyber Assets to support reliable operation of the Bulk Electric System.
IEEE 1686-2007	IEEE Standard for Substation Intelligent Electronic Devices (IEDs) Cybersecurity Capabilities.
ISO/IEC 27001:2005	Information technology—Security techniques—Information security management systems—Requirements.
	Can handle distributed generation securely.
ANSI/ISA-99	Security for Industrial Automation and Control Systems.

The December 2015 Ukraine power blackout helps as a demonstrative sample of cyberattacks against the power grid. In this incident, hackers acquired control of the human–machine interfaces at three Ukrainian power plants and affected a blackout lasting for about 10 h and upsetting over 225,000 people. The post operations conducted by a joint task force of the US FBI, Department of Homeland Security ICS-CERT team, and Ukrainian authorities determined that the corresponding attacks initially targeted susceptible software in the IT operations and detected Black Energy malware that navigated through internal networks within minutes to attack the control systems. Power was reestablished rather quickly by disconnecting the computer systems and manually resuming the systems (IR-ALERT-H-16-056-01., 2016).

NIST also recommends the cybersecurity as a most important factor because SG information is very important and it should be secured for efficient operation of SG. For SG applications, there are three important cybersecurity objectives, confidentiality, availability, and integrity. The ideal trade-offs are necessary for balancing the most important aspects of SG information. The standards of communication efficiency and information security should not be compromised in the design of communications protocols and architectures as both are important considerations for the SG operations. The SG infrastructure incorporates millions of intelligent electronics devices across the network. For secure operations across SG network, the strict implementation for recognizing the device (authentication), its validation, and access control becomes more important. For this, it is very important that every node in SG should have standard cryptography functions and should perform better data encryption. Utmost care should be taken to ensure that every node in SG is secure from cyberattacks, for this, network operations should perform profiling testing regularly (Wang & Lu, 2013).

The cyberattack on SG communications and security challenges is one of priority researches that strengthens the overall performance of SG operations. There are two types of security attacks, selfish misbehaving users, and malicious misbehaving users. Selfish misbehaving users try to obtain more network resources than authentic users by breaching communication protocols (Pelechrinis et al., 2009). Second, malicious misbehaving attacks shall persuade disastrous damage to the power supplies results in the power outage, which is prescribed in SG.

The malicious attacks are of three types that affect the accessibility, reliability, and confidentiality of SG objectives. The attacks affecting the accessibility are also called as denial-of-service (DoS) attacks that mainly interrupt, block, or alter the SG communications. Attacks targeting the reliability aim for unlawful modifications or disrupting of the data communication in the SG. Attacks targeting confidentiality are intentional in obtaining the unauthorized information from network resources in the SG (Wang & Lu , 2013).

The SG networking identified "channel-jamming" threat at PHY layer of networking protocol (Lu et al., 2011); this type of attack can cause the wide range of damages to the local area networks affecting the network performance, particularly, the substations and home areas by delaying the packet delivery time for critical messages to complete denial-of-service. The threat "Spoofing" is identified at the MAC layer where an attacker can take an advantage of the address fields in MAC frames and can mask that address to attacker node to communicate the fake information to other devices (Premaratne et al., 2010).

It is very important to focus on the attacks on the information and communication networks of SG. Signal-based detection is carried at PHY layer or MAC layer where a DoS attacker can detect the presence of attack based on the received signal strength indicator (RSSI) information. If the RSSI of several data packets is greater than a threshold (that means the receiver node should properly receive them). But if the packet decoder at receiver records errors in the received data, the attack detector can raise an alarm in the presence of an attacker (Lu et al., 2014). Some jamming-resistant protocols include UFHSS, UDSSS, and UFH-UDSS (Popper et al., 2010), DEEJAM (Wood et al., 2007), Timing-channel (TC) (Xu et al., 2008), and JADE (Richa et al., 2010). Packet-based detection solutions can be implemented at any of the layers of SG networking and can compare packet delivery ratio at regular time intervals if significant packet transmission failures are detected then it serves as an alarm for attacks (Toledo et al., 2008).

In consideration of all the aspects above, a secure SG should achieve the following security goals, data confidentiality, message authentication, inter message sequencing and deletion, repetitive message detection, privacy preserving, and revocation. The SG networking applications require the standard solutions designed specifically for distinctive network applications that are making the cyber security a challenging research for the future.

9.5 CONCLUDING REMARKS

With the advent of modern technologies such as information, communication technology (ICT) the traditional grid is getting transformed to the SG with increased advanced automation and becoming hope for future energy needs. The most important factor for considering the SG is efficient energy management with the introduction of distributed generation of energy with the association of renewable energy sources. The DR is handled logically, and peak loads can be shifted or postponed for other times. This way of managing the electricity shall not only increase the stability of the system but also decreases the carbon emissions and protects the environment.

The most important infrastructure for SG is establishing AMI technology for introducing the bidirectional communication to learn about the electric usage between the end user and utilities, thereby introducing dynamic tariff with active participation at end user level. AMI includes the technologies such as WAN, NAN, and HAN for communication purposes. Different communication technologies serve different purposes based on the types of applications and work with various data rates. The interoperability issues among technologies such as communication, information, and data management need to be critically addressed and there is a lot of scope for future research.

Maintenance of power quality is one of the major concerns addressed by SG. The smart meter at end user premises and intelligent electronic devices (IEDs) at distribution systems shall manage the voltage levels and power factor. The smart meters shall record the voltage levels delivered at end user premises and informs the utility center about this information at regular time intervals. With this data utility centers shall optimize the voltage levels thereby increase the power quality of the system. With the better voltage, the appliances at the end user premises work with higher efficiency.

In the power systems scenario, situation-based operation or event-based operations is not sufficient for controlling and cannot guarantee the system stability. The remote monitoring is the most prime advantage of SG technology. The remote monitoring devices include distribution transformers, capacitor banks, phasor measurement units, smart meters, etc. The concepts of remote monitoring and wide area monitoring have generated technologies WAMS and WAMPAC in the support of SG for enhanced management of power losses, faults, and disturbances. These monitoring technologies will lower the power outages, increases the power delivery, decreases operational costs and increases the end user satisfaction.

With the increased use of renewable energy resources such as solar energy at end user premises shall benefit the customer to tackle with real-time pricing for DR billing, reduced energy bills, and do better planning for load usage or peak shifting. Vehicle-to-grid and gird-to-vehicle charging is becoming the prime concern with better utilization of energy resources because of their high impact on the power systems, hence, high research is to be focused on this area by industry and academia.

The main concerns pertaining utilities such as efficient data management, controlling, and communication of information needs to be concerned for efficient SG operations. The privacy issues related to huge data integrity and confidentiality are the prime concern of SG needs to be focused upon. The utilization of communication technologies increases the interconnections among various appliances of SG that introduces the vulnerabilities because of cyber security and related issues, which need to be addressed by fine-grained technical approaches based on SG issues.

KEYWORDS

- **smart grid**
- **energy management**
- **advanced metering infrastructure**
- **communication networks**
- **cyber security**

REFERENCES

Amin, S. M. Smart Grid: Overview, Issues and Opportunities. Advances and Challenges in Sensing, Modeling, Simulation, Optimization and Control. *Eur. J. Cont.* **2011**, 5(6), 547–567.

Amin, M.; Stringer, J. The Electric Power Grid: Today and Tomorrow. *MRS Bull* **2008**, 33(4), 399–407.

Anda, M.; Temmen, J. Smart Metering for Residential Energy Efficiency: The Use of Community Based Social Marketing for Behavioural Change and Smart Grid Introduction. *Renew. Energy* **2014**, 67, 119–127.

Bartoli, A.; Hernandez-Serrano, J.; Soriano, M.; Dohler, M.; Kountouris, A.; Barthel, D. In *Secure Lossless Aggregation for Smart Grid m2m Networks*, First IEEE International Conference on Smart Grid Communications, Gaithersburg, USA, 2010, 333–338.

Bayindir, R.; Colak, I.; Fulli, G.; Demirtas, K. Smart Grid Technologies and Applications. *Renew. Sustainable Energy Rev.* **2016**, 66, 499–516.

Cyber-Attack Against Ukrainian Critical Infrastructure. Alert (IR-ALERT-H-16-056-01). https://ics-cert.us-cert.gov/alerts/IR-ALERT-H-16-056-01

Emmanuel, M.; Rayudu, R. Communication Technologies for Smart Grid Applications: A Survey. *J. Network Comput. Appl.* **2016**, 76, 133–148.

Ericsson, G. N. Cyber Security and Power System Communication—Essential Parts of a Smart Grid Infrastructure. *IEEE Trans. Power Delivery* **2010,** *25,* 1501–1507.

Expert Group on the Security and Resilience of Communication Networks and Information Systems for Smart Grids. Cyber Security of the Smart Grids. Summary Report. European Commission. 2012.

Gao, W.; Jin, W.; Li, H. In *An Improved Routing Protocol for Power-line Network Based on AODV,* 11th International Symposium on Communications and Information Technologies (ISCIT), 2011, 233–237.

Gao, J.; Xiao, Y.; Liu, J.; Liang, W.; Philip Chen, C. L. A Survey of Communication/ Networking in Smart Grids. *Future Gener. Comput. Sys.* **2012,** *28*(2), 391–404.

Gharavi, H.; Hu, B. Multigate Communication Network for Smart Grid. *Proc. IEEE 99,* **2011,** 1028–1045.

Gomez, C.; Paradells, J. Wireless Home Automation Networks: A Survey of Architectures and Technologies. *IEEE Commun. Mag.* **2010,** *48*, 92–101.

Haider, H. T.; See, O. H.; Elmenreich, W. A Review of Residential Demand Response of Smart Grid. *Renew. Sustainable Energy Rev.* **2016,** *59*, 166–178.

Han, D. M.; Lim, J. H. Smart Home Energy Management System Using IEEE 802.15.4 and Zigbee. *IEEE Trans. Consumer Electron.* **2010,** *56*(3), 1403–1410.

Heo, J.; Lee, K.; Kang, H. K.; Kim, D. -S.; Kwon, W. H. Adaptive Channel State Routing for Home Network Systems Using Power Line Communications. *IEEE Trans. Consumer Electron.* **2007,** *53*, 1410–1418.

Ho, Q.-D.; Gao, Y. Rajalingham, G.; Le-Ngoc, T. Smart Grid Communications Network (SGCN). Wireless Communications Networks for the Smart Grid. Springer, 2014.

Hossain, E.; Han, Z.; Vincent Poor, H. Smart Grid Communications and Networking. *Cambridge University Press.* 2012.

Hossain, A.; Chakrabarti, S.; Biswas, P. K. Impact of Sensing Model on Wireless Sensor Network Coverage. *IET Wireless Sensor Syst.* **2012,** *2*(3), 272–281.

IEEE Standards Coordinating Committee 21. IEEE Std 2030. IEEE Guide for Smart Grid Interoperability of Energy Technology and Information Technology Operation with the Electric Power System (EPS), End-use Applications, and Loads. 2011.

Iwao, T.; Yamada, K.; Yura, M.; Nakaya, Y.; Cardenas, A.; Lee, S.; Masuoka, R. In *Dynamic Data Forwarding in Wireless Mesh Networks,* First IEEE International Conference on Smart Grid Communications (SmartGridComm.), 2010, 385–390.

Jarrah, M.; Jaradat, M.; Jararweh, Y.; Al-Ayyoub, M.; Bousselham, A. A Hierarchical Optimization Model for Energy Data Flow in Smart Grid Power Systems. *Inf. Sys.* **2015,** *53*, 190–200.

Jin, C.; Kunz, T. In *Smart Home Networking: Combining Wireless and Powerline Networking;* 7th International Wireless Communications and Mobile Computing Conference (IWCMC), 2011, 1276–1281.

Juanjuan, W.; Chuang, F.; Yao, Z. Design of WAMS-based Multiple HVDC Damping Control System. *IEEE Trans. Smart Grid.* **2011,** *2*(2), 363–374.

Jung, J.-S., et al. In *Improving IEEE 802.11s Wireless Mesh Networks for Reliable Routing in the Smart Grid Infrastructure.* 2011 International Conference on Communications Workshops (ICC), 2011, 1–5.

Keyhani, A. Design of Smart Power Grid Renewable Energy Systems; *John Wiley & Sons, Inc.* 2011.

Kuzlu, M; Pipattanasomporn, M.; Rahman, S. Communication Network Requirements for Major Smart Grid Applications in HAN, NAN and WAN. In *Computer Networks*; 2014; Vol. 67, pp. 74–88.

Li, F.; Luo, B.; Liu, P. In *Secure Information Aggregation for Smart Grids Using Homomorphic Encryption*. 2010 First IEEE International Conference on Smart Grid Communications. Gaithersburg, MD, 2010, 327–332.

Lu, Z.; Wang, W.; Wang, C. In *From Jammer to Gambler: Modeling and Detection of Jamming Attacks Against Time-critical Traffic*, Proc. of IEEE INFOCOM 2011, 2011.

Lu, Z.; Wang, W.; Wang, C. Modeling, Evaluation and Detection of Jamming Attacks in Time-critical Wireless Applications. *IEEE Trans. Mobile Comput.* **2014**, *3*(8), 1746–1759.

Ma, R.; Chen, H. H.; Huang, Y. R.; Meng, W. Smart Grid Communication: Its Challenges and Opportunities. *IEEE Trans. Smart Grid* **2013**, *4*(1), 36–46.

Mahmood, A.; Javaid, N.; Razzaq, S. A Review of Wireless Communications for Smart Grid. *Renew. Sustain. Energy Rev.* **2015**, *41*, 248–260.

Meliopoulos, P. S.; Polymeneas, E.; Tan, Z.; Huang, R.; Zhao, D. Advanced Distribution Management System. *IEEE Trans. Smart Grid.* **2013**, *4*(4), 2109–2117.

Meng, W.; Ma, R.; Chen, H. H. Smart Grid Neighborhood Area Networks: A Survey. *IEEE Netw.* **2014**, *28*(1), 24–32.

Mets, K.; Ojea, J. A.; Develder, C. Combining Power and Communication Network Simulation for Cost-effective Smart Grid Analysis. *IEEE Commun. Surveys Tutorials* **2014**, *16*(3), 1771–1796.

Mikhail, G. Catching the Z-wave. *Electronic Engineering Times India.* **2006**, 1–5.

Mohassel, R. R.; Fung, A.; Mohammadi, F.; Raahemifar, K. A Survey on Advanced Metering Infrastructure. *Electr. Power Energy Sys.* **2014**, *63*, 473–484.

Mouftah, H. T.; Erol-Kantarci, M. Smart Grid: Networking, Data Management, and Business Models. *CRC Press.* 2016.

Nafi, N. S.; Ahmed, K.; Gregory, M. A.; Datta, M. A Survey of Smart Grid Architectures, Applications, Benefits and Standardization. *J. Network Comput. Appl.* **2016**, *76*, 23–36.

Namboodiri, V.; Aravinthan, V.; Jewell, W. Communication Needs and Integration Options for AMI in the Smart Grid. *White Paper. PSERC Publication.* 2012.

NIST Special Publication 1108, NIST Framework and Roadmap for Smart Grid Interoperability Standards, Release 1.0. January 2010.

Paper, W. Wireless Wan for the Smart Grid, *Trilliant Inc.,* **2010**.

Pathak, V. Meter Data Acquisition System (MDAS) Implementation Challenges in India R-APDRP. *Smart Energy Metering Int. Issue* **2013**.

Pelechrinis, K.; Yan, G.; Eidenbenz, S. In *Detecting Selfish Exploitation of Carrier Sensing in 802.11 Networks*, Proc. of the IEEE Conference on Computer Communications (INFOCOM '09), 2009.

Petersen, S.; Carlsen, S. Wirelesshart Versus isa100.11a: The Format War Hits the Factory Floor. *IEEE Ind. Elect. Mag.* **2011**, *5*(4), 23–34.

Popper, C.; Strasser, M.; Capkun, S. Anti-jamming Broadcast Communication Using Uncoordinated Spread Spectrum Techniques. *IEEE J. Sel. Areas Commun.* **2010**, *28*(5) 703–715.

Premaratne, U.; Samarabandu, J.; Sidhu, T.; Beresh, R.; Tan, J.-C. An Intrusion Detection System for IEC61850 Automated Substations. *IEEE Trans. Power Deliv.* **2010**, *25*, 2376–2383.

Ho, Q.-D.; Gao, Y.; Rajalingham, G.; Le-Ngoc, T. Wireless Communications Networks for the Smart Grid. Springer Publishing Company, Incorporated, 2014.

Richa, A.; Scheideler, C.; Schmid, S.; Zhang, J. In *A Jamming-resistant MAC Protocol for Multi-hop Wireless Networks*. Proceedings 24th International Symposium, DISC 2010, Cambridge, MA, USA, Sept 13–15, 2010, 179–193.

Saputro, N.; Akkaya, K.; Uludag, S. A Survey of Routing Protocols for Smart Grid Communications. *Comput. Netw.* **2012**, *56*(11), 2742–2771.

Shomali, A.; Pinkse, J. The Consequences of Smart Grids for the Business Model of Electricity Firms. *J. Clean. Prod.* **2016**, *112*(5), 3830–3841.

SMB Smart Grid Strategic Group (SG3). IEC Smart Grid Standardization Roadmap. 2010.

Terzija, V., et al. Wide-area Monitoring, Protection, and Control of Future Electric Power Networks. *Proc. IEEE* **2011**, *99*, 80–93.

Toledo, A. L.; Wang, X. Robust Detection of MAC Layer Denial-of Service Attacks in CSMA/CA Wireless Networks, *IEEE Trans. Inf. Forensics Secur.* **2008**, *3*, 347–358.

Uribe-Pérez, N.; Hernández, L.; de la Vega, D.; Angulo, I. State of the Art and Trends Review of Smart Metering in Electricity Grids. *Appl. Sci. MDPI.* **2016**, *6*, 68, 1–24.

Vikram, K.; Venkata Lakshmi Narayana, K. In *Cross-layer Multi Channel MAC protocol for Wireless Sensor Networks in 2.4-GHz ISM Band*. IEEE Conference on, Computing, Analytics and Security Trends (CAST-2016). at Department of Computer Engineering & Information Technology, College of Engineering, Savitribhai Phule Pune University, Pune, Maharashtra, India. Dec 19–21, 2016.

Wall, P.; Dattaray, P.; Jin, Z., et al. Deployment and Demonstration of Wide Area Monitoring System in Power System of Great Britain, *J. Mod. Power Sys. Clean Energy* **2016**, *4*(3), 506–518.

Wang, W.; Lu, Z. Cyber Security in the Smart Grid: Survey and Challenges. *Comput. Netw.* **2013**, *57*(5), 1344–1371.

Wireless Medium Access Control (MAC) and Physical Layer (PHY) Specifications for Low-rate Wireless Personal Area Networks (WPANs) Amendment 4: Physical Layer Specifications for Low Data Rate Wireless Smart Metering Utility Networks. IEEE Std. P802.15.4g/D4 Part 15.4, Apr 2011.

Wood, A. D.; Stankovic, J. A.; Zhou, G. DEEJAM: In *Defeating Energy-efficient Jamming in IEEE 802.15.4-based Wireless Networks*, Proc. of IEEE SECON '07, 2007, 60–69.

Xi, F.; Satyajayant, M.; Guoliang, X.; Dejun, Y. Smart Grid—the New and Improved Power Grid: A Survey. *IEEE Commun. Surv.* Tutorials **2012**, *14*(4), 944–980.

Xu, W.; Trappe, W.; Zhang, Y. In *Anti-jamming Timing Channels for Wireless Networks*, Proc. of ACM Conference on Wireless Security (WiSec), 2008, 203–213.

Yu, R.; Zhang, Y.; Gjessing, S.; Yuen, C.; Xie, S.; Guizani, M. Cognitive Radio Based Hierarchical Communications Infrastructure for Smart Grid. *IEEE Netw.* **2011**, *25*(5), 6–14.

Yu, Y.; Liu, Y.; Qin, C. Basic Ideas of the Smart Grid. *Engineering* **2015**, *1*(4), 405–408.

CHAPTER 10

SMART-GRID INTERACTION WITH ELECTRIC VEHICLES

RAMJI TIWARI[1] and N. RAMESH BABU[2*]

[1]*School of Electrical Engineering, VIT University, Vellore, Tamil Nadu 632014, India*

[2]*M. Kumarasamy College of Engineering, Karur, Tamil Nadu, India*

[*]*Corresponding author. E-mail: nrameshme@gmail.com*

CONTENTS

ABSTRACT

The transportation sector is the one of the most dominating sectors for the fuel consumption which in turn increases the carbon emission. Transportation is one of the most needful sectors for any nation. Thus, constant depletion of fossil fuel gives an alarm to find a suitable alternative to the transportation. The electric vehicle is suitable and more efficient alternative for the fuel-based vehicle. The constant increase in fuel prices urges for one to opt for electric vehicle which also helps to reduce the carbon emission caused due to burning of fuel. In this chapter, the importance of vehicle to grid (V2G) concept is discussed. The integration of renewable energy like solar and wind is analysed along with the strength and weakness of the V2G concept. The need of the energy storage system and battery storage system is also presented to provide the detail analysis of the electric vehicle.

10.1   INTRODUCTION

The fossil-fuel-based energy generation is more dominant in present world for both transportation sector and power sector. The faster depletion of fossil fuel and constant growth in demand gives an alarming call to find an alternative source for both sectors. The oil economy in the world is highly dependable on the vehicle fleet; thus, shortage of oil reserve in future will bring many vehicles to halt. In addition to the shortage, burning of the fossil fuel produces the greenhouse gases and carbon emission which is a global challenge for climatic change. According to report (Mwasilu et al., 2014), the oil consumption in transport sector will increase by 54% until year 2035 as the population is also rising at a higher rate. The rise in consumption in parallel with decrease of the oil reserve tends to broaden the gap between supply and demand, thus increasing the cost of oil manifold in near future.

In context to reduce the oil consumption, various researches related to development of alternative energy in transport and power generation sector has been carried out in near future. Electric vehicles (EV) are one of such remarkable solution, which not only eliminates the use of oil but also helps to reduce the carbon emission. The electrification of transportation sector will be one of the feasible solutions for challenges like global warming, energy security, reliability, and avoiding the natural disturbance to extract the oil from sea. EV also can be used as the storage system which can deliver the power back to grid using vehicle to grid (V2G) technology.

The EV technology can provide the grid support by charging the battery during the low peak time and then deliver the stored energy back to grid in peak time. EV acts as a portable distributed energy source which has a potential to deliver power when ever needed. High penetration of renewable energy into grid requires a large amount of energy storage system (ESS) to support the grid, since the renewable energy sources (RES) are intermittent in nature. Forecast of RES are also almost unpredictable. The EV act as storage devices which provide an additional support to the electrical demands and to always meet the operational standard of the power grid. Virtual power plant concept model is used to control the EV when the V2G context is used to provide aggregated and reliable power (Vasirani et al., 2013).

Though the EV system has major benefits to world in transport sector, in power sector, they may face many challenges like charging and discharging of battery packs, power-system grid reliability, operation, and control of grid. The operating cost and maintenance cost of EV are other challenges which have to be reduced for more implementation. Smart charging/discharging units should be installed so that the time taken to charging and discharging can be reduced without any power loss (Galus et al., 2012).

The contemporary penetration of EV in the grid requires an effective solution to calculate the cost associate in charging the EV. Thus, a dynamic pricing of the electricity should be introduced so that the consumers can charge the vehicle during low-energy demand where the prices will be low and sell the power by delivering it into grid in high-peak demand when the cost is high so as to earn some profit. The real-time pricing or dynamic pricing is quite intuitive and requires high sophisticated and advance metering devices like advanced metering infrastructure (AMI) (Ortega-Vazquez et al., 2013).

The EV is considered as the dynamic load which can be shifted as required in the future electrical grid or smart grid. The important parameter of smart grid which differs from the existing grid is a two-way communication which requires information and communication technology (ICT), the data are stored using cloud computing and hence requires a security to avoid misleading of the information (Gungor et al., 2013).

This chapter provides an overview of electric vehicle's role in development of smart grid in Section 10.2. The integration of EV with the renewable sources, like solar and wind, is also discussed in Section 10.3. In Section 10.4, impact of V2G concept in smart grid with its strength and challenges is discussed.

10.2   ELECTRIC VEHICLE INTEGRATION WITH A SMART GRID

The smart grid encompasses advanced technologies, smart metering, ICTs, and advanced controls which enhance the flexibility, reliability, and security of the power at generation and distribution end. The EV is one of such examples where the flexibility of smart grid is determined. The EV can act as a dynamic load when the demand of electricity is low and delivers power to grid as a source during peak demand. Many studies have been carried out in the literature to access and realize the smart-grid infrastructure to optimize the EV penetration into grid (Su et al., 2012). Standardization of communication protocol of electric power distribution is a key parameter behind the implementation of interactive smart grid. Each country which uses EV to interact with the grid has their own protocol and standard specification that should be strictly followed.

The EV also uses communication system to send and receive information from the grid operator. The real-time energy measurement, efficient communication, and advance control in EV can be achieved by embedding smart meter. The availability of power in grid with the real-time price of power can be optimized using smart scheduling through the bidirectional exchange of the data.

Smart charging is the primary and most important parameter of EV which controls the undesirable impact of power during charging. Smart meters are essential to charge the EV rapidly and also help to minimize the cost of charging. Optimized and smart price algorithms are also implemented in the past (Mal et al., 2013) which facilitates intelligent charging and uses radiofrequency identification tag to involve all EV users with a specific and electronic identity. The information of real-time statistic, control of EV charging by using parameters like state of charge, and statistic about consumed and delivered power is connected through Web-mobile-based application. The scheduled charging scheme is a cost-effective solution where the consumers can save 10% of cost and time when compared with flexible charging scheme. In addition, the scheduled scheme has a reduction of peak demand of about 54% than variable scheme (Mwasilu et al., 2014).

Energy management system (EMS) in smart grid is performed by measuring the data, analyzing the data based on the total energy used, and total demand of energy in near real time. To perform EMS in efficient process, smart-metering technology is used. In the integration of EV in the smart grid, smart meter plays an important role in providing information regarding the total power consumed and demanded. Hence, the forecast of the data can be nearly estimated which can provide more feasible pricing information. Advanced metering information is a framework which can accommodate dynamic EV loads and provide real-time smart metering and bidirectional communication in single device. Meter data management, advanced sensor network which is installed in different substations and communication technologies are incorporated in AMI which can provide all the data of real time and can be retrieved when required. The communication networks in AMI can be wireless or broadband power line which provides faster and secure two-way communications between utility, consumers, smart meters, sensors, and EV management system. The information stored by AMI can be used to optimize to implement intelligent decision and control system. Thus, it can be stated that deployment of EV using AMI platform can manage the EV-charging scenario, information of power consumed and demanded which helps to reduce peak energy demand and shift the energy in off-peak period. Hence, the efficient power management is performed which releases the stress of power system (Lun et al., 2011).

Wireless communication is an effective solution for EV which can gather the data on the go. Wireless communication is an advanced communication network which features low-cost data transfer over a wide area. In EV, interaction with smart meters or with smart grid requires frequent request of real-time data for successful operation. The communication architecture of EV is carried out in two different scenarios,

1) communication of smart meters with EV and sensors and
2) communication of smart meters with the data centers and grid operators.

The communication technology for the first scenario can be of broadband lines or wireless, whereas the second scenario uses mobile network solutions like 3G, WiMAX, and 4G (Mwasilu et al., 2014).

The dynamic nature of EV when deployed in the power industry brought newer challenges in monitoring, communicating, and controlling of power and data. The advanced smart meters should allow the EV to communicate with other grid operators when they are away from their home or local network to use the smart management system anywhere and anytime. Wireless sensor is another promising application for inter-action of EV with the grid operators which enables them to charge the EV anytime and deliver the power to the grid when required with real-time pricing option. Figure 10.1 shows the communication interaction of EV with the smart grid. The architecture shows the employment of smart phone as an interface portal for the EV with the charging station, manage-ment systems, grid operators, and data centers which is enabled using GPS or Bluetooth function. To increase the reliability of EV interaction with the smart grid, automatic Bluetooth pairing using near-field communication protocol is used (Steffen et al., 2010). Thus, the smart-grid interaction with EV provides efficient solution to reduce the cost of electricity and support the grid in high peak demand when an efficient and secure communication network is established.

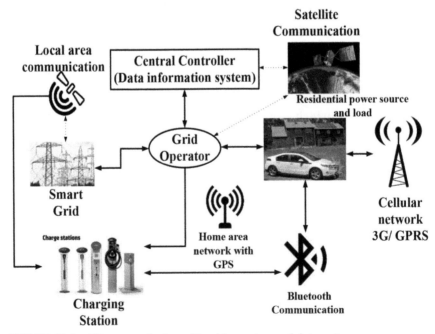

FIGURE 10.1   EV communication with grid operators and data centers.

10.3   ELECTRIC VEHICLE INTEGRATION WITH RENEWABLE ENERGY

The penetration of renewable sources such as solar and wind into electric grid is rapidly increasing throughout the world. The RES are intermittent in nature especially the solar and wind energy. The power production in these sources may increases and decrease irrespective to the energy demand based on the availability of the solar radiation or wind sources. Thus, integration of these unpredictable sources to the grid is not an effective way to overcome the energy demand. The EV is a promising solution to balance the energy generated from these RES and energy delivered to grid by acting as an ESS and as load when the power is in excess. The stationary energy storage plants require high investment cost and serve only one purpose, whereas EV serves two purposes, one in power storing and another is the transportation; both purposes are helpful to eradicate the issue of climatic change (Mwasilu et al., 2014). The EV acts as a dynamic and portable ESS. The charging and discharging of EV is based on the availability of RES and load demand; thus, this may reduce the need of storage system for RES which reduces the cost of system. The interaction of EV and RES has an added advantage of reducing the carbon emission when fuel-powered cars are used. Figure 10.2 shows the integration of RES with the EV and grid.

The electricity production from PV solar energy already has shown a feasible alternative in recent time. Generally, the PV panels are connected in cluster to provide a supply to the grid. At the time when a large number of electric vehicles are used, the PV panels can be used to charge the battery system of EV. The PV panels deployed in the residential by the consumers can also use the electricity produced from it to charge the EV and then later deliver it to the grid in return of cash or incentives. Roof-top PV panels may reduce the need of charging station in the country and the users can charge their vehicle any time when the solar irradiation is high without focusing much on the peak demand of the system. The PV panels deliver DC power which can be used to charge the EV just by using a DC–DC converter that has greater efficiency when compared to other converters. Thus, integration of solar with the EV is very simple and feasible for every consumer (Tulpule et al., 2013).

The wind energy conversion system (WECS), which is termed as high nonlinear system, requires an efficient solution to deliver the generated power to the grid. Formerly, WECS system uses many reactive power and fault ride through mechanism to fulfill the grid standards. But these techniques increase the complexity and cost of the system. Implementation of wind energy using EV in integration with the grid reduces the cost and complexity of WECS by eliminating many controllers in the system. Thus, even a small power produced from the wind can be delivered in the grid using EV which increases the total power production of wind. EV requires an additional converters and charge controllers to protect the battery from being damaged since the wind power continuously vary with time. The advance controllers which are used to optimize the wind output power are also eliminated which reduces the overall implementation cost of WECS (Wu et al., 2013).

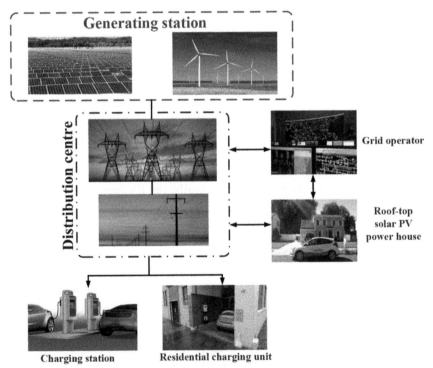

FIGURE 10.2   Integration of renewable energy with electric vehicle.

Thus, in spite of being intermittent in nature, both solar and wind energy can deliver all the power generated by them to grid without using advanced and complex controllers. Thus, the overall efficiency of the RES is increased and cost and complexity of the system is drastically reduced. The RES can be used as the charging station which reduces the implementation of charging stations everywhere and RES can provide an additional source of income for the users.

10.4 V2G: IMPACT, POTENTIAL, AND CHALLENGES

V2G system is a technique which enables to communicate with the grid either by delivering the power or by utilizing the grid power to charge their battery. V2G system is used as the dynamic distributed energy source that can provide a support to the electrical grid in peak time. The typical battery management system, which is used to store the energy, requires huge investment to the power producers. But EV with V2G terminology is user centric where the user spends money to buy EV and then later support the grid during peak demand at huge rates, whereas the charging would have been taken in off-peak demand when the price of the electricity is low. Thus, this serves as an additional income to an individual. V2G also serves as a very efficient alternative to the transportation. As they are battery operated, the need of fuel is not required except for some hybrid electric vehicles which have both fuel- and battery-based engine. Thus, cost of fuel is saved which is lot more when compared to the electricity and no or less use of fuel limit in emission of carbon which is the major threat to global warming and climatic change. The V2G system is divided into three different concepts (Lund and Kempton, 2008):

i) **Hybrid vehicle:** The fuel cell is used to generate the power and the vehicle serves as distribution system. Conventional fossil fuel, biofuel, or hydrogen is used as the base material to generate the power which is later transferred to grid when there is peak demand.

ii) **Plug-in hybrid vehicle:** Plug-in hybrid vehicle uses rechargeable battery as the storage device. They store the power during off-peak demand at cheaper rates and deliver the power when there is peak-demand at higher rates. This kind of vehicle serves as energy storage plants.

iii) **Solar vehicle:** Solar vehicle is associated with the direct implementation of solar power to charge the vehicle when there is excess power generated and low power demand. And in some case, the excess power generated after charging the battery is transferred to grid. This kind of vehicle serves as small renewable power generation.

The V2G systems are mainly classified into two types, namely unidirectional V2G and bidirectional V2G. The system which only delivers the power into grid and does not utilize the grid power to charge the vehicle is termed as unidirectional V2G. The bidirectional charges and discharges the power for grid to vehicle and V2G, respectively, based on the energy demands. These types of system mainly contribute in auxiliary control of electrical grid by efficient utilizing the excess power generated (Sortomme and El-Sharkawi, 2012).

Though the V2G concept has a potential, there exist many challenges in implementation of this technology in real time. The major challenge which is associated with this concept is proposer synchronization and coordination of several EV which can act as a single storage device. Since the EV can accommodate only small battery backup, several EV is required to provide sufficient power to the grid when there is peak demand. Thus, this can be only possible when government take appropriate measure to increase the interest of EV among the people by providing subsidies. The next challenge is the proper communication. The communication in V2G is very essential because it can only provide the current real-time status of power-grid demand and the price of electricity. The battery warranty is also very essential because continuously charging and discharging power to the grid reduces the lifetime of the battery which incur an additional cost to the users. The last challenge is providing enough information about V2G among the peoples. Lack of knowledge makes the effective system also fail and lose interest in that. Thus, the V2G system has a very good potential in near future which will increase the grid reliability and 100% electricity throughout the world.

CONCLUSION

In this chapter, the interaction of electric vehicle with the smart grid is discussed. The integration of renewable energy and electric vehicle is also discussed. The electric vehicles can provide ancillary support to grid in peak demand, regulating the voltage and frequency, and reactive power support. They act as dynamic load in off-peak demand which utilizes the additional power produced by the RES. Thus, an additional control component in the renewable-based generation is reduced which provides cheap electricity. The need of advanced charging, metering, and communication in smart-grid-based electric vehicle was also discussed. The feasibility of V2G in smart grid has been explored which plays an important role in energy management system.

KEYWORDS

- smart grid
- electric vehicles
- renewable energy
- vehicle to grid (V2G)
- advance metering infrastructure (AMI)

REFERENCES

Galus, M. D.; Vayá, M. G.; Krause, T.; Andersson, G. The Role of Electric Vehicles in Smart Grids. *Wiley Interdiscip. Rev. Energy Environ.* **2013,** *2*, 384–400.

Gungor, V. C., et al. A Survey on Smart Grid Potential Applications and Communication Requirements. *IEEE Trans. Ind. Inform.* **2013,** *9*, 28–42.

Lun, K. L., et al. Advanced Metering Infrastructure for Electric Vehicle Charging. *Smart Grid Renew. Energy* **2011,** *2*, 312–323.

Lund, H.; Kempton, W. Integration of Renewable Energy into the Transport and Electricity Sectors through V2G. *Energy Policy* **2008,** *36*, 3578–3587.

Mal, S.; Chattopadhyay, A.; Yang, A.; Gadh, R. Electric Vehicle Smart Charging and Vehicle-to-Grid Operation. *Int. J. Parallel Emerg. Distrib. Syst.* **2013,** *28*, 249–265.

Mwasilu, F., et al. Electric Vehicles and Smart Grid Interaction: A Review on Vehicle to Grid and Renewable Energy Sources Integration. *Renew. Sustain. Energy Rev.* **2014,** *34*, 501–516.

Ortega-Vazquez, M. A.; Bouffard, F.; Silva, V. Electric Vehicle Aggregator/System Operator Coordination for Charging Scheduling and Services Procurement. *IEEE Trans. Power Syst.* **2013,** *28*, 1806–1815.

Sortomme, E.; El-Sharkawi, M. A. Optimal Combined Bidding of Vehicle-to-Grid Ancillary Services. *IEEE Trans. Smart Grid* **2012,** *3*, 70–79.

Steffen, R.; et al. Near Field Communication (NFC) in an Automotive Environment. In *Int. Workshop Near-Field Commun.* 2010; pp 15–20.

Su, W.; Eichi, H.; Zeng, W.; Chow, M. Y. A Survey on the Electrification of Transportation in a Smart Grid Environment. *IEEE Trans. Ind. Inform.* **2012,** *8,* 1–10.

Tulpule, P. J.; Marano, V.; Yurkovich, S.; Rizzoni, G. Economic and Environmental Impacts of a PV Powered Workplace Parking Garage Charging Station. *Appl. Energy* **2013,** *108,* 323–332.

Vasirani, M., et al. An Agent-Based Approach to Virtual Power Plants of Wind Power Generators and Electric Vehicles. *IEEE Trans. Smart Grid* **2013,** *4,* 1314–1322.

Wu, T.; Yang, Q.; Bao, Z.; Yan, W. Coordinated Energy Dispatching in Microgrid with Wind Power Generation and Plug-in Electric Vehicles. *IEEE Trans. Smart Grid* **2013,** *4,* 1453–1463.

CHAPTER 11

SMART GRIDS: GLOBAL STATUS

RAMJI TIWARI[1] and N. RAMESH BABU[2*]

[1]*School of Electrical Engineering, VIT University, Vellore 632014, Tamil Nadu, India*

[2]*M. Kumarasamy College of Engineering, Karur, Tamil Nadu, India*

Corresponding author. E-mail: nrameshme@gmail.com

CONTENTS

ABSTRACT

Smart grid is coined as the future power system network which is able to manage the entire power system network and provides two-way communications with the end users. Smart grid is an efficient tool which provides effective utilization of generated power which in turn reduces the production cost. The concept of smart grid is growing exponentially for continuous advancement in technologies. The need of the smart power management system is must since more renewable energy sources are connected to the distribution system. In this chapter, a study based on the challenges faced by the smart grid interfacing is presented along with the future research perceptive of the smart grid. The worldwide status of the smart grid implementation is also adapted in this chapter. The technologies such as advance metering infrastructure, information and communication technologies are also described to provide an overlook of the smart grid system.

11.1   INTRODUCTION

Smart grid is considered as the future power grid which manages the power production, transmission, and distribution using modern and advance technology to overcome the technical issues such as power quality and uninterrupted power supply in current grid system. The problems in grid are classified as economical and environmental. The obstacles of economical problems are voltage sag, overload, current leakage, and blackouts. The factors such as global warming and carbon emission contribute the environmental issue. The environmental issue can be addressed using the renewable energy as the source to produce power. To limit the economical issue, modern control strategy, advance transmission device, communication devices, and cloud computing are used. Thus, application of smart grid implemented with renewable energy source (RES) will be essential in near future. Modernization of power grid by implementing communication and data mining system is rapidly emerging throughout the world. Moreover, high penetration of renewable energy in power grid increases the risk of intermittent power supply. Thus, smart grid is essential to balance the load based on the usage providing uninterrupted power (Phuangpornpitak & Tia, 2013).

The key goal of smart grid is to provide effective communication between customer and provider which enhances the reliability of power based on the utility. Demand response, energy storage at substation, and automated grid system with self-healing capability provide consistent power with low cost (Buchholz & Styczynski, 2014).

The implementation of smart grid and integration of RES in smart grid face many technical challenges due to many issues such as lack of knowledge among customers, cost, and irrelevant load pattern. As renewable energy is uncertain, the power system reliability is also a major concern. The other nontechnical challenges are policy, grid investment, data access, and grid security (Cecati et al., 2011).

This chapter provides an overview of challenges faced by smart grid and future research perspective in smart grid in Section 11.2. The case study of renewable energy in the world along with their role in smart grid is briefly explained in Section 11.3. Section 11.4 provides comprehensive remarks of the chapter for betterment of near future.

11.2 CHALLENGES AND RESEARCH PERSPECTIVES IN SMART GRID

The carbon emission in the environment and energy efficiency are two major threats which motivate the implementation of smart grid. Smart grid manages the production of power, transmission of power, and consumption of power in the customer end to ensure the efficiency, stability, security, and affordability of power (Fadaeenejad et al., 2014). The basic architecture of smart grid is shown in Figure 11.1. There are other architectures which encompass the nonrenewable source and other distribution system. But Figure 11.1 shows a universally accepted model of smart grid integrated with the renewable source which has two-way communications with data flow. The reliability of smart grid operation depends on the proper communication and data protection (Fan et al., 2013).

There exists a significant challenge in integration of smart grid with the renewable system. Traditionally, power grid has an issue of uncertainty, addressing load demand. But with the integration of renewable energy, the intermittency of supply side should also be managed. Smart grid requires direct load control, utility side load control, and a mechanism to adjust the consumption of power directly and indirectly (Samadi et al., 2012).

Thus, making awareness among the consumers about the load pattern should be practiced. The indirect mechanism to adjust the consumption of power such as providing incentives to the consumers for shifting their load usage is already in practice in many countries. Dynamic pricing mechanism overcomes the instability of supply–demand gap during peak hours (Gungor et al., 2011).

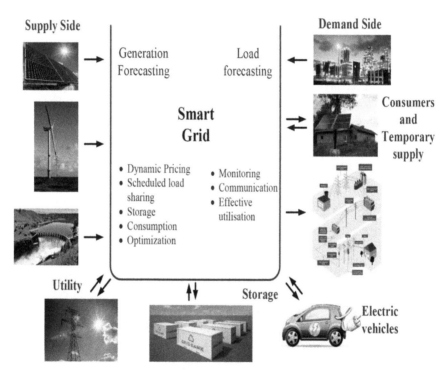

FIGURE 11.1   Architecture of smart grid.

As defined earlier, the integration of renewable, which is inconsistent in nature, requires additional storage such as battery, fuel cell hybrid electric vehicles (HEV) or other distributed generation to overcome the demand. The distributed generation manages the varying grid supply by supporting the local demands. The storage devices store the power when in excess or in low demand time which can improve the stability of the system during peak demand (Moslehi & Kumar, 2010). The HEVs are major load and source unit. This system can be charged during low demand time mostly overnight and delivered when there is power shortage. HEV can be profitable when there exists dynamic pricing. The vehicle to grid (V2G) technology is playing a vital role, overlapping the issue of smart grids. The charging of HEV overnight also ensures that the generated power is optimally utilized because the wind power produces more power during night when there is no peak demand (Ban et al., 2012).

The major research area to be focused in smart grids is providing effective and secure communication between service provider and consumer. Advanced metering infrastructure (AMI) and data storage are imposed on a communication network to provide a fast and secure connection between the supply side and the load side (Farhangi, 2010). The challenge faced in this area is the generation of a control signal that provides information of different consumption costs at different times. Customer behavior plays a vital role in deciding the parameters, such as profitability, usability, stability, and load ability requirement of power. Multiobjective problems are derived from various large problems to reduce the complexity of the system, which tend to further determine load profiles of the consumer end.

Another promising area to be focused in smart grid technology is the control of power electronics. Almost all RESs, storage devices, and HEVs use power electronic components to integrate with grid. Power electronic interface is used to eliminate the disturbance during grid variations. The power electronic control alleviates the problems of grid frequency and the voltage collapse. The control of power electronic is a vast research perceptive in terms of smart grid to overcome the undesirable effect of renewable energy within the bounds of physical capabilities (Amin & Wollenberg, 2005).

The energy forecasting is very much essential to predict the estimated number of consumer load demand at a particular time interval and making the users to participate in demand management in return of incentives. Short-term load forecasting is done for single or monolithic load in demand response. Multiple short-term load forecasting are done for a single load to anticipate its accuracy. To forecast many loads, short-term multiple load forecasting is performed which reduces the scalability problem of single short-term forecasting method. It is proved that short-term multiple load forecasting has an accuracy of 7% higher than that of former method (Potter et al., 2009).

Thus, smart grid can incorporate many distributed RESs to cope up with energy demands. The plug and play integrations of generator radially improve the smart grid capability, so as to regulate the voltage and add many larger RESs. In order to enhance the reliability of smart grid, frequency control and responsive load which can be altered based on the demand is introduced. Thus, by integrating high distribution renewable sources, the power outage problem is minimized (Petinrin & Shaaban, 2012). Despite of these features, there exists a challenge of synchronization of many small RESs which are integrated with the grid. Issues such as harmonics, fluctuation, and ripple in generated voltage are caused. Thus, the smart grid should enhance the communication network to allow the user to efficiently manage their energy utilization, and storage facility should be added to overcome the power shortage during peak demand. The system should be flexible enough to add any number of source and storage in the transmission and distribution system (Mohd et al., 2008).

11.3 SMART GRID AND RENEWABLE ENERGY STATUS WORLDWIDE

11.3.1 RENEWABLE ENERGY STATUS

The global community has started to adopt the renewable energy as the primary source of power in order to address the issue of global warming and reduce the production of greenhouse gases due to burning of fossil fuel. Utilization of electrical energy is the key growth of any country economically. Particularly, the developing nations are very much keen to increase the renewable-based generation, as it is abundantly available and also acts as a cheap source of electricity (Camacho et al., 2011). Huge investments are made by various countries in developing and implementing the renewable energy-based power production. Renewable energy also plays a vital role is electrifying the rural and remote areas where the transmission of grid power is impossible. The major renewable sources which produce the electricity are solar, wind, hydro, tides, geothermal, and biomass. Figure 11.2 shows the general production of electricity at present, worldwide. The share of renewable energy in the overall power production is shown in Figure 11.3 (REN, 2016).

Based on REN21's (a global renewable energy stakeholder network) 2016 report, 19.2% of global energy consumptions by users are generated by renewable energy (REN, 2016). The worldwide investment of US$ 286 billion has been made for development of renewable energy. Table 11.1 provides share of different RESs which have been installed worldwide from 1849 GW of total renewable capacity as of 2015. Table 11.2 determines the top five countries in total investment made by them in developing various renewable energies.

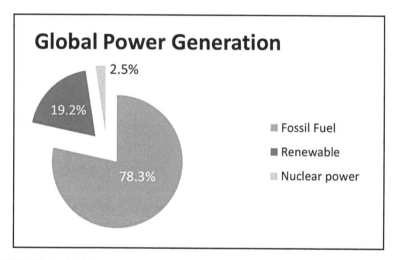

FIGURE 11.2 Global production of electricity.

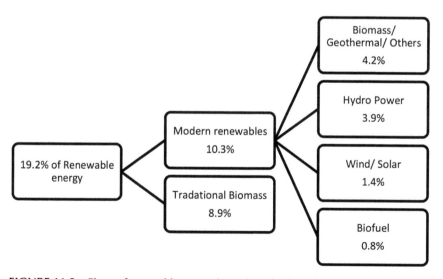

FIGURE 11.3 Share of renewable energy in total production of power.

TABLE 11.1   Total Share of Different Renewable Energy Installed Worldwide (2015).

Renewable energy	Installed capacity
Solar PV	227 GW
Wind power	433 GW
Hydro power	1064 GW
Biopower	106 GW
Geothermal	13.2 GW
Others (ethanol, heat—solar …, etc.)	5.8 GW

TABLE 11.2   Top Five Countries Based on Their Annual Investment in Developing Renewable Energy (as of End-2015).

Ranking/renewable energy	1	2	3	4	5
Total investment	China	United States	Japan	United Kingdom	India
Solar PV	China	Japan	United States	United Kingdom	India
Wind power	China	United States	Germany	Brazil	India
Hydro power	China	Brazil	Turkey	India	Vietnam
Biopower	United States	Brazil	Germany	Argentina	France
Geothermal	Turkey	United States	Mexico	Kenya	Japan

From the Table 11.1, it can be observed that the hydro-based power production contributes almost 70% of total renewable-based generation. The main reason behind high production is that the water being 800 times denser than air. Thus, a small stream flowing water can also generate considerable amount of electricity. The major hydropower generation is based on dams. There exist many dams which have capacity to generate 50 MW of power. China is the largest producers of hydroelectric power, followed by Brazil and Venezuela. Wave power and tidal power are the other forms of hydro-based power generation techniques which capture the kinetic energy of wave and then transform into electricity. However, these techniques are not yet widely commercialized, except for some small pilot project to monitor their feasibility. The other method to harness the electricity from ocean is ocean thermal energy conversion which utilizes the temperature difference between the cooler and warmer region of water, but the economic feasibility is still not clear. World's highest tidal-based generation plant is located in the Bay of Fundy, which is a pilot project operated by Ocean Renewable Power Company. This plant is also connected to grid (Karsten et al., 2008). Table 11.3 shows the top five countries with the installed renewable energy sources.

Wind power contributes 4% of total electricity demand in the world. Wind energy is one of the most promising RESs which has highest ratio in terms of development in developing nations. Wind energy meets the majority of the total electricity demands in the countries such as Denmark, Spain, and Portugal. The long-term technical potential of wind energy is believed to be five times of total current global energy production. China stands topmost in power production, also using wind energy. Offshore wind turbines are gaining a lot more interest due to its high and continuous wind speed throughout the year. The largest country which has commissioned the offshore wind power is United Kingdom, managing almost 1.3 GW of offshore wind energy. Siemens and Vestas contribute 90% of total offshore wind power installed (GWEC GW, 2016).

Solar power contributes 1% of total energy production globally. The solar energy is also used and installed widely. The major focus on the implementation of solar energy is to develop an affordable, inexhaustible, and clean energy. It has huge long-term benefit with one time high investment. The developing countries require secure and indigenous power which can be overcome using solar PV plants. Solar-based alternative energy is also highly booming which uses heat to convert to another form of energy (PS REN, 2016).

The geothermal energy extracts the thermal energy which is stored in the earth crust. The geothermal energy utilizes the heat and temperature present in the earth to generate steam which is then used to run the power plant and produce electricity. Thus, this reduces the burning of coal which is basically used to heat the water to generate the steam. The geothermal is advancing at the rate of 5% every 3 years. According to Geothermal Energy Association (GEA), only 6.5% of total potential of geothermal has been tapped. The countries such as Kenya, Philippines, Iceland, Costa Rica, and El Salvador are generating more than 15% of geothermal energy of their total energy production. United States ranked first in total installed capacity of geothermal which generates 28% of total geothermal-based power production (Dye, 2012).

TABLE 11.3 Top Five Countries Based on Their Installed Capacity of Different Renewable Energy (as of End-2015).

Ranking/Renewable energy	1	2	3	4	5
Total installed capacity (including hydro)	China	United States	Brazil	Germany	Canada
Total installed capacity (without hydro)	China	United States	Germany	Japan	India
Solar PV	China	Germany	Japan	United States	Italy
	(43.53 GW)	(39.700 GW)	(34.41 GW)	(25.62 GW)	(18.92 GW)
(178 GW)					
Wind power	China	United States	Germany	India	Spain
	(145.362 GW)	(74.471 GW)	(44.947 GW)	(25.088 GW)	(23.025 GW)
(433 GW)					
Hydro power	China	United States	Brazil	Canada	Russia
	(311 GW)	(102 GW)	(89 GW)	(76 GW)	(51 GW)
(2848 GW)					
Biopower	United States	China	Germany	Brazil	Japan
	(15.4 GW)	(3.71 GW)	(1.1 GW)	(0.8 GW)	(0.433 GW)
(35 GW)2010					
Geothermal	United States	Philippines	Indonesia	Mexico	New Zealand
	(3.45 GW)	(1.87 GW)	(1.340 GW)	(1.017 GW)	(1.005 GW)
(12.8 GW)					

The biomass-based energy is also referred as bioenergy. The biomass indirectly contributes in the prevention of electricity in another form. The biomass is an organic material which stores the sunlight in the form of chemical energy. The bioenergy plant uses the by-products of organic material such as sugarcane, wood, manure, and many more as a fuel to produce electricity (Gonzalez-Salazar et al., 2016). The bioenergy power plants have an installed capacity of 35 GW, as of 2010. United States has a highest installed capacity of 15.4 GW. The other countries which have installed bioenergy-based power plants are Brazil (28%), Europe (16.5%), and Asia (10.6%) (IEA, 2015).

11.3.2 SMART GRID—GLOBAL STATUS

The term smart grid generally refers to an electric grid which can have a two-way communication between consumers and producers. The important objective of the smart grid is to develop an advanced electricity infrastructure with sophisticated communication, precise control, and high reliability (Irfan et al., 2016). The concept of smart grid was implemented in early 2000. Many achievements in the field of smart grid were developed by many countries. Each country has their own objective definition according to their requirement (Fadaeenejad et al., 2014). The approach on smart grid is different in different countries. This section provides an overview of major countries involved in developing the smart grid technology based on their objectives and legislation.

11.3.2.1 CHINA

China is one of the largest producers of renewable energy. The objectives of smart grid in China focus on three main areas. The main objective of smart grid in China is efficient and clean energy.

i)   Generation expansion to meet the high demand,
ii) Transmission and distribution of generated electricity, and
iii) Reducing carbon emission.

China being highly industrially productive country with high electricity demand requires more power generation plants to fulfill the generation needs. China plans to build new coal plants to expand their generation. As stated above, China has put greater effort in development of renewable energy. The largest expansion will be in hydroelectric power plants as they have vast exposure to ocean. The expansion of China's generation capacity has been estimated to 120 GW from hydroelectric, 70 GW from wind energy, and solar will contribute 5 GW of installed capacity by 2020. Nuclear power plant in China also has a greater expansion of about 40 GW. Ultrahigh voltage (UHV) transmission lines are to be installed to connect the generating station with the grid. UHV lines transfer the power in low cost and minimal loss. The new coal plants which are to be installed in China will follow clean coal technology (CCT) to reduce the carbon emission by burning them.

China smart grid technology will be more transmission centric. Wide area monitoring system (WAMS) technology is implemented by China to monitor the generation and transmission of China network. PMU sensors are set up in all power plants which have generating capacity of 300 MW and above. The substations which have capacity of 500 kV are also installed with the PMU sensors. All the communication in the generation and transmission are adhered to their standard uses broadband to deliver the data without any delay over private network (12th 5-year plan of China) (Den et al., 2016).

11.3.2.2 UNITED STATES

Smart grid technology in United States has funding of about $100 million in 2012. The funding is used to establish a modern grid with smart meters and secure communication between utility and consumers. Protocol standards are recommended using the benefits of demand response. The smart grid standards in the United States are developed by National Institute of Standards and Technology. The Federal Energy Regulatory Commission (FERC) issues an official policy statement and action plans.

Smart grid technology in the United States is governed by FERC. The main objectives of smart grid in the United States are clean energy, demand response system, energy storage system, and electric vehicle. The electric vehicle is deployed to be charged during low power demand and it can be integrated with V2G capability which can act as a distributed generation and energy storage system.

The smart grid with secure communication and reliability is the primary focus in the United States. The United States has allocated more funding than any other country for implementation of smart grid. The entire electrical grid in the United States will be installed with 850 sensors which will make the grid operators to monitor the grid condition and allow them to alter their load based on the availability of renewable energy. Seven hundred automated substations will also make the power producers to respond faster during natural calamities or any electricity disruption. The Unites States will install 2.5 million smart meters within 2020 which makes the consumer to access the dynamic pricing and to avoid expensive pricing during peak hours. The smart grid technology in the United States can also better accommodate use of plug-in electric vehicle and power produced by the consumer which can be transferred to grid (Tomain, 2016).

11.3.2.3   INDIA

Indian grid is termed as the weakest grid in the world. India looses 26% of total power generated during transmission and distribution. Power theft is also a major concern in India. If power theft is included, then around 50% of total power generated in India is lost during transmission. Indian grid lacks in poor plan of distribution system, reactive power management, and overloading of system component. Pricing and metering efficiency in India is also low.

Economically, India is also gaining a lot more interest in renewable-based generation. Solar and wind power based generation have a greater potential in India. With such a highly developing nation with huge number of integration of renewable energy generation, existing grid is not suitable since they are associated with lot of losses in power and economy. Thus, India requires a highly adaptive grid which can balance the supply and demand. Building of smart grid in India is more important because efficient electric supply is a key infrastructure for overall development.

Bangalore Electricity Supply Company (BESCOM) is working on pilot projects of implementation in India. The key objectives of smart grid implementation in India are power quality, minimal losses in transmission and distribution, reliability, power efficiency, and customer friendly cum satisfaction. Power grid in India requires more advance mechanism to achieve the goal of 100% electricity. Proper monitoring of distribution

centers and power theft is required. The energy planning, huge investment, and awareness about smart grid are some important parameters to
be considered for implementation of smart grid in India (Fadaeenejad et
al., 2014).

11.3.2.4   EUROPE

European Commission named European technology platform (ETP) is the
initiative of European electricity community for the development of smart
grid. European grid unification was achieved parallel to the economical
unification of European countries. Modernization of European grid is
based on mainly control, automation, and monitoring. The key objectives
of European smart grid are the distributed generation, storage devices, and
power electronics. Demand-side management is also considered as a key
parameter in smart grid implementation in European grid. Demand-side
management is targeted to achieve energy efficiency and dynamic pricing.
Several dynamic pricing methods have been implemented to achieve the
better reflection of power production cost and the incentive given to the
consumers.

Plug-in hybrid electric vehicles (PHEV) are also used to overcome the
energy demand during peak times. Data management in European smart
grid is still in critical situation, since they need to collect, process, validate, and transmit huge amount of raw data. Protection and automation
of distribution system is an advantage of European smart grid. Thus, this
provides continuous, reliable, and secure supply of electricity. Automation
in home and industries is a key infrastructure of development of smart
grid in Europe which enhances two-way communication and self-healing
capability (Iqtiyanilhan et al., 2017).

11.3.2.5   AUSTRALIA

Australian government also has a keen interest in developing smart grid.
They have promised to invest $100 million in near future. The federal
government has proposed a bid to study about smart grid technology. The
study is intended to create awareness among people about smart grid,
energy usage profile, load management, and study on distributed generation management. Australia has made a consortium with leading software

companies including IBM, Grid net, and Energy Company such as GE to build a smart grid over five cities. The WiMAX-based smart grid will support the automation of substations. PHEV are used to support for the cause. 50,000 smart meters are to be installed with 15,000 in-home devices (IHD). The major hindrance in smart grid implementation in Australia is lack of service level obligation on distribution generations to connect with the consumers in modern and advanced method (Zhang, 2016).

11.3.2.6 BRAZIL

Brazil will be one of the most potential markets for smart grid for its growing economy and high investment in infrastructure. Siemens has invested $1 billion in smart grid development in Brazil for next 5 years. Thus, smart grid infrastructure can be forecasted until 2022. Brazil is primarily focused on their objective of generation of electricity using renewable energy and enhancing the grid infrastructure. The driving forces for the development of smart grid in Brazil are,

i)   high demand of electricity,
ii) reliable on hydroelectric power, and
iii) high nontechnical losses.

Brazil is continuously investing in development of renewable energy. The high reliability of Brazil in renewable energy requires an advance grid infrastructure to overcome the intermittency. Biomass and hydroelectric plants have much higher market in Brazil than that of solar.

Brazil is planning to install the smart meter to all new customers and in rural places, where renewable energy is a main source of electricity. They have an option for existing customers where they can request for smart meter. Brazil has estimated to install 27 million smart meters by end of 2030. The nontechnical losses which are high in Brazil can be reduced by installing the smart meters (Fadaeenejad et al., 2014).

11.4 CONCLUSION

In this chapter, an overview of current status of renewable energy installed and smart grid utilization in different countries has been discussed.

Developing countries such as China, India, and Brazil are the main market of smart grid and renewable energy implementation. This chapter also provides the challenges and research opportunities in smart grid as well as in renewable implementation. The fact that can be concluded is that the developing countries are very much keen in implementation of smart grid for their future development. The high rise in addition of renewable energy may reduce the risk of global warming. Integration of renewable energy is a key factor for development of smart grid. The smart grid may reduce the losses and increase the chances of demand management. The benefits and the challenges of smart grid with renewable energy are also discussed in this chapter.

KEYWORDS

- smart grid
- renewable energy
- global report of installed renewable energy
- smart grid with renewable energy
- electric vehicles

REFERENCES

Amin, S. M.; Wollenberg, B. F. Toward a Smart Grid: Power Delivery for the 21st Century. *IEEE Power Energy Mag.* **2005,** *3,* 34–41.

Ban, D.; Michailidis, G.; Devetsikiotis, M. In *Demand Response Control for PHEV Charging Stations by Dynamic Price Adjustments,* Proc. IEEE PES Innovative Smart Grid Technologies (ISGT), 2012; pp 1–8.

Buchholz, B. M.; Styczynski, Z. *Smart Grids—Fundamentals and Technologies in Electricity Networks*; Springer, 2014.

Camacho, E. F., et al. Control for Renewable Energy and Smart Grids. In *The Impact of Control Technology*; Control Syst. Society. 2011; pp 69–88.

Cecati, C.; Citro, C.; Siano, P. Combined Operations of Renewable Energy Systems and Responsive Demand in a Smart Grid. *IEEE Trans. Sustain. Energy* **2011,** *2*(4), 468–476.

Den, E. M., et al. Greenhouse Gas Emissions from Current and Enhanced Policies of China Until 2030: Can Emissions Peak Before 2030? *Energy Policy* **2016,** *89,* 224–236.

Dye, S. T. Geoneutrinos and the Radioactive Power of the Earth. *Rev. Geophys.* **2012,** *50.*

Fadaeenejad, M., et al. The Present and Future of Smart Power Grid in Developing Countries. *Renew. Sustain. Energy Rev*. **2014**, *29*, 828–834.

Fan, Z., et al. Smart Grid Communications: Overview of Research Challenges, Solutions, and Standardization Activities. *IEEE Comm. Surveys Tutor*. **2013**, *15*, 21–38.

Farhangi, H. The Path of the Smart Grid. *IEEE Power Energy. Mag*. **2010**, *8*, 18–28.

Gonzalez-Salazar, M. A., et al. Development of a Technology Roadmap for Bioenergy Exploitation Including Biofuels, Waste-to-energy and Power Generation & CHP. *Appl. Energy* **2016**, *180*, 338–352.

Gungor, V. C., et al. Smart Grid Technologies: Communication Technologies and Standards. *IEEE Trans. Ind. Inform*. **2011**, *7*, 529–539.

GWEC. 2016. http://gwec.net/global-figures/wind-energy-global-status/ (accessed Oct 20, 2016).

International Energy Agency: Bioenergy (IEA). 2015, https://www.iea.org/topics/renewables/subtopics/bioenergy/ (accessed Oct 20, 2016).

Iqtiyanilham, N.; Hasanuzzaman, M.; Hosenuzzaman, M. European Smart Grid Prospects, Policies, and Challenges. *Renew. Sustain. Energy Rev.* **2017**, *67*, 776–790.

Irfan, M., et al. Opportunities and Challenges in Control of Smart Grids—Pakistani Perspective. *Renew. Sustain. Energy Rev*. **2016** (Article in Press).

Karsten, R. H., et al. Assessment of Tidal Current Energy in the Minas Passage, Bay of Fundy. *Proc. Inst. Mech. Eng. A J. Power Energy* **2008**, *222*, 493–507.

Mohd, A., et al. In *Challenges in Integrating Distributed Energy Storage Systems into Future Smart Grid*, *Proc. Int. Symp. Ind. Electron.* 2008; pp 1627–1632.

Moslehi, K.; Kumar, R. A Reliability Perspective of the Smart Grid. *IEEE Trans. Smart Grid* **2010**, *1*, 57–64.

Petinrin, J. P.; Shaaban, M. S. Overcoming Challenges of Renewable Energy on Future Smart Grid. *TELKOMNIKA* **2012**, *10*, 229–234.

Phuangpornpitak, N.; Tia, S. Opportunities and Challenges of Integrating Renewable Energy in Smart Grid System. *Energy Procedia* **2013**, *34*, 282–290.

Potter, C. W.; Archambault, A.; Westrick, K. In *Building a Smarter Smart Grid Through Better Renewable Energy Information*, Proc. Power Syst. Conf. Expos., 2009; pp 1–5.

REN. *Renewables 2016 Global Status Report*; REN21 Secretariat: Paris, France, 2016.

Samadi, P., et al. Advanced Demand Side Management for the Future Smart Grid Using Mechanism Design. *IEEE Trans. Smart Grid* **2012**, *3*, 1170–1180.

Tomain, J. P. A Perspective on Clean Power and the Future of US Energy Politics and Policy. *Util. Policy* **2016**, *39*, 5–12.

Zhang, X. Secure Data Management and Transmission Infrastructure for the Future Smart Grid. Sydney E-scholarship Library, 2016.

INDEX